T0213006

# Progress in IS

More information about this series at http://www.springer.com/series/10440

Arnold Picot · Thomas Hess
Christian Hörndlein · Natalie Kaltenecker
Claudius Jablonka · Michel Schreiner
Alexander Werbik · Alexander Benlian
Rahild Neuburger · Bernhard Gold

# The Internationalization of German Software-based Companies

## Sustainable Growth Strategies for Small and Medium-sized Companies

 Springer

Arnold Picot
Research Center for Information, Organization
and Management
Ludwig-Maximilians-Universität München
Munich
Germany

Thomas Hess
Institute for Information Systems and New Media
Ludwig-Maximilians-Universität München
Munich
Germany

Christian Hörndlein
Institute for Information Systems and New Media
Ludwig-Maximilians-Universität München
Munich
Germany

Natalie Kaltenecker
Institute for Information Systems and New Media
Ludwig-Maximilians-Universität München
Munich
Germany

Claudius Jablonka
Center for Digital Technology and Management
Munich
Germany

Michel Schreiner
Institute for Information Systems and New Media
Ludwig-Maximilians-Universität München
Munich
Germany

Alexander Werbik
Research Center for Information, Organization
and Management
Ludwig-Maximilians-Universität München
Munich
Germany

Alexander Benlian
Information Systems and Electronic Services
Technische Universität Darmstadt
Darmstadt
Germany

Rahild Neuburger
Research Center for Information, Organization
and Management
Ludwig-Maximilians-Universität München
Munich
Germany

Bernhard Gold
T-Venture of America, Inc
San Francisco, CA
USA

ISSN 2196-8705          ISSN 2196-8713   (electronic)
Progress in IS
ISBN 978-3-319-38054-4     ISBN 978-3-319-13548-9   (eBook)
DOI 10.1007/978-3-319-13548-9

Springer Cham Heidelberg New York Dordrecht London
© Springer International Publishing Switzerland 2015
Softcover reprint of the hardcover 1st edition 2015

Printed on acid-free paper

Springer International Publishing AG Switzerland is part of Springer Science+Business Media
(www.springer.com)

# Acknowledgments

We would like to thank all of the German and international experts, interview and case study partners, as well as the participants of our empirical survey, who took the time to support us in the course of this scientific study. We appreciate their effort, without which this study would not have been possible.

We are especially thankful to Ingo Ruhmann from the Federal Ministry of Education and Research and Dr. Jens Totz from the German Aerospace Center for their constructive feedback throughout all phases of the project.

Last but not least, we would like to thank Dr. Rebecca Ermecke and Dr. Kathrin Sattler for their expertise and methodological support in the different phases of the project. We also thank all of the student research assistants for the support and dedication throughout the whole project: Alysa Reuter, Beate Ströhlein, Christian Soyk, Claudia Beinert, Corinne Petroschke, Heinrich Rusche, Johanna von Boeselager, Leonie Schminke, Maria Farber, Markus Zuleger, Michaela Rauscher, and Moritz Palme.

# Contents

# Chapter 1
# Introduction

The software industry is central to Germany's economic competitiveness. Software-based companies play a crucial role in the competition between the economies of the 21st century. They are critical both as a key industry on its own and functioning as a driver for many other industries and sectors, such as the automotive industry (e.g. embedded systems), engineering (e.g. Industry 4.0), the energy sector (e.g. Smart Energy), or medical devices (e.g. e-health). The software industry's importance will continue to grow, as machines are increasingly being integrated with traditional software components to form autonomously acting cyberphysical systems. Considering the economy's transformation into a digital economy, it is of uttermost importance that Germany catches up with the international leaders in the most important areas.

The global market for business-to-consumer (B2C) software for operation systems (e.g. Microsoft, Apple), social networks (e.g. Facebook), social news (e.g. Twitter), user-generated content (e.g. YouTube), search engines (e.g. Google), as well as retail and marketplaces (e.g. Amazon, eBay) is currently dominated by US-American companies. US-American companies, such as Oracle, Microsoft, and IBM have also taken the lead in the business-to-business (B2B) segment. In contrast, only two of the world's largest software companies, SAP AG and Software AG, were founded in Germany. A large number of German companies primarily operate in specialized market segments.

Compared to the dominating companies in the software industry, software-based companies from Germany usually play a minor role. There have been some individual successes in the recent past, where German companies could grow to be international leaders in certain market segments, such as the gaming sector (Söndermann 2010, p. 28). One could also observe an increase in start-up and entrepreneurial activity. Nevertheless, the German market is still dominated by small and very small software development companies with a low export share (Söndermann 2010, p. 28, 93).

This development is problematic for the following two reasons: on the one hand, software-based companies' importance will continue to grow, as the economy will

© Springer International Publishing Switzerland 2015
A. Picot et al., *The Internationalization of German Software-based Companies*,
Progress in IS, DOI 10.1007/978-3-319-13548-9_1

be increasingly digitalized (Broy et al. 2006, p. 212). On the other hand, software-based companies will face a stronger pressure to grow and internationalize their business in the face of globally distributed value chains. In such a setting, the ability to scale up one's business activities will be a critical competitive advantage.

Considering the background that we have laid out, Germany is facing the issue of how small and medium-sized software-based companies can successfully grow and internationalize their business to gain traction in the global market. However, dealing with this challenge cannot be left to the public sector alone. Entrepreneurs and managers of local software-based companies need to understand which strategies are likely to be successful, given the conditions they face in Germany. Just imitating the successful US-American ecosystem *"Silicon Valley"* is unlikely to be successful. Instead, one has to address the question regarding the specific measures that German companies can take to gain a position among the world's leading companies.

This question was addressed by the research project *"DESC—Deutsche Software-Champions (German Software Champions): Status quo, success factors, perspectives"*, which was funded by Germany's Federal Ministry of Education and Research. The results of this research project are summarized in this research report. The focus was placed on German small and medium-sized software-based companies.

The research project's objective was to identify the conditions on a macro-economic level that are especially relevant for the growth and internationalization of software-based companies. On a business level, this project set out to derive successful internationalization strategies.

These two levels were approached from different perspectives and analyzed through expert interviews, a large empirical survey among small and medium-sized software-based companies, as well as national and international cases studies with selected companies.

Based on the differentiated results on a business and macro-economic level, we could develop recommendations for managers and the public sector.[1]

We were able to point out measures of how to improve the macro-economic conditions that would increase software-based companies' chances to internationalize successfully. In addition, we were able to identify the success factors of how ambitious software-based companies can become international leaders within the existing conditions and constraints they currently operate in.

This study offers entrepreneurs and managers in German-speaking countries specific hints in regards to promising internationalization strategies, and presents them with best practices that were identified among the "champions" of this study's software-based companies. The study also points out to the public sector certain conditions and factors in the ecosystem that can accelerate the growth of software-based companies.

---

[1] In this study, we refer to the government and trade/industry associations as the public sector.

The project was conducted between January 2011 and June 2013 by a team of several scientists from the Munich School of Management at the Ludwig-Maximilians-University (LMU) Munich. The project was led by Prof. Dr. Dres. h.c. Arnold Picot (Research Center for Information, Organization, and Management[2]) and Prof. Dr. Thomas Hess (Institute for Information Systems and New Media[3]). It was conducted in close cooperation with the Center for Digital Technology and Management (CDTM[4]), a joint institution of the LMU and the Technical University Munich (TUM).

Regular presentations of intermediate results, e.g. at the national IT summit 2011 and 2012, and a steady exchange with representatives from relevant associations (e.g. BITMi and BITKOM) ensured that the remarks and suggestions of subject matter experts and specialists accompanied the entire project, which were then incorporated directly into the study.

---

[2] http://www.iom.bwl.uni-muenchen.de/wegweiser.

[3] http://www.en.wim.bwl.uni-muenchen.de.

[4] http://cdtm.de/.

# Chapter 2
# Motivation

## 2.1 Status Quo of the German and Global Software Industry

Recent studies have shown that there is strong growth in the software industry. The global market for software and software services grew by 7.8 % in 2012 (OECD 2012, p. 55). The industry association BITKOM was expecting a market growth of 3.1 percent for 2013. This means that Germany was predicted to reach a market volume for software and IT services of 53.6 billion euros in 2013 (BITKOM 2013).

The export of software products and services from Germany amounts to 12.1 billion euros, about half of which (6.1 billion euros) is exported to European countries (BITKOM 2011). The two largest German software companies, SAP AG and Software AG, which respectively generate 82.4 and 79.5 % of their revenue abroad (Lünendonk 2011), contributed the highest share.

At the same time, German companies only sporadically take a leading role on the international stage. There are 15 Germany companies among Europe's 100 largest software companies, accounting for 50 % of the top 100 companies' revenue. However, excluding SAP, Germany only accounts for 8 % of the top 100 European companies' revenue (Truffle Capital 2012). The global picture is even more discouraging: while only three of the world's top 100 software companies are from Germany, 63 companies are from the USA (Top 100 Research Foundation 2011). Germany and Europe are also underrepresented when one considers the world's top 50 publicly traded Internet companies. Germany with only 2 companies (corresponding to 4 %) and the rest of Europe with five companies (corresponding to 10 %) have a far lower percentage than the USA, which dominate the ranking with 31 US-American Internet companies (corresponding to 62 %; OECD 2012, p. 44). Therefore, one can conclude that compared to the size of the entire economy, the Germany software industry is underrepresented on a global scale.

There is a large gap between the high number of very small companies and the small number of large (at least for German standards) companies in the software

© Springer International Publishing Switzerland 2015
A. Picot et al., *The Internationalization of German Software-based Companies*,
Progress in IS, DOI 10.1007/978-3-319-13548-9_2

and IT services industry. This indicates that the weakness of the Germany software industry is not primarily due to the start-up phase, but can rather be found in the later growth phase (cf. Scheer 2001 on this topic). The concentration on a small number of larger companies and on the other hand a lack of medium-sized companies provides evidence that there are impediments in the growth trajectory of software companies (Leimbach 2010, p. 10). This imbalanced structure also impacts exports, as large companies, in particular, are export-oriented, whereas small and young companies exhibit considerable deficits in their export orientation and internationalization (Lünendonk 2011). This can become problematic as the software and IT services market is very international and the quick diffusion of innovative solutions can be seen as a crucial competitive advantage (Leimbach 2010, p. 27).

## 2.2 Terminology

The project set out to explore how to successfully foster the growth and internationalization of software-based companies. Therefore, we will first provide our understanding of the terms "software-based companies" and "internationalization" and how these terms are used within this study.

### 2.2.1 Software-Based Companies

Software-based companies offer products and services with software at its core. In addition to traditional software companies that develop software either for individual clients or for the mass market, this definition of software-based companies also includes web-centric companies and embedded systems companies. We chose this extended definition as their underlying technological and economic mechanisms are similar or even identical. Our study did not focus on pure IT service providers. Although their business is related to software, they are, as customer-focused service providers, subject to different economic principles. Therefore, their offerings are more similar to traditional service companies than to scalable software development companies (Cusumano 2004, p. 26).

Despite, or rather because their common basis is software, which can be used for a plethora of purposes, software-based companies cover a variety of different products and business models. Their offerings range from apps to standard application software to custom-developed enterprise software. We developed a matrix to identify differences within the different segments. For this purpose, we distinguish between the functional domain (system software, application software, web-centric applications, embedded systems, and other) and the level of the value chain (development and sales of own software, implementation and hosting, as well as consulting and training). The matrix therefore also includes web-centric applications and embedded systems, which are not part of the software industry in a narrow sense.

We did not focus on companies whose main source of revenue is based on implementing and hosting software or on consulting and training for the purpose of this study. The following "DESC software matrix" (cf. Table 2.1) was used specifically for classifying companies in the quantitative survey.

Certain characteristics apply to the market in which software-based companies operate. These characteristics refer to both the software itself and the market environment (Buxmann et al. 2011). While immateriality and scalability are characteristic of the software itself, software markets are coined to some extent by strong network effects. In addition, technological advances and disruptions play an important role.

**Table 2.1** DESC software matrix

| Main revenue source of the responding companies[a]: | | | |
|---|---|---|---|
| | Development and sales of own software | Implementation and hosting | Consulting and training |
| System software: | | | |
| Operating systems | o | o | o |
| Security software | o | o | o |
| Database | o | o | o |
| Other system software | o | o | o |
| Application software: | | | |
| Business application software (e.g. ERP) | o | o | o |
| Technical application software (e.g. CAD) | o | o | o |
| Games | o | o | o |
| Other application software | o | o | o |
| Web-centric applications: | | | |
| Internet portals | o | o | o |
| Knowledge platforms (e.g. rating platforms) | o | o | o |
| E-commerce / market places of all sorts | o | o | o |
| Social networks (incl. dating platforms) | o | o | o |
| Search engines | o | o | o |
| Mobile applications (e.g. apps, navigation) | o | o | o |
| Online advertising | o | o | o |
| Other web-centric applications | o | o | o |
| Embedded systems | o | o | o |
| Other | | o | |

[a] The survey was conducted without color coding the different categories

*Dark gray* software-based companies in a narrow sense
*Light gray* software-based companies in a broad sense
*White* not within the main focus of this study

The conversion into binary formats and the resulting immateriality reduce the necessity of an integration into physical value chains and distributional structures (e.g. Picot and Neuburger 2010), thus enabling the replication of perfect copies at variable costs approaching zero (Buxmann et al. 2011). This makes it possible to benefit from economies of scale. Scalability in this context refers to low variable, respectively low marginal, costs for every additional copy of digital products, as well as decreasing unit costs for services, which make it easier to market offerings at flexible prices. Both factors impede the market entrance of competitors and create sustainable profits for dominant companies.

Moreover, software markets are characterized by network effects. In such "winner-takes-it-all" markets (Buxmann et al. 2011), direct network effects (Zerdick et al. 2000) are created when additional users generate a higher utility for existing users. Typical examples are platforms such as Facebook, eBay, or Twitter. Strong network effects in combination with large scale effects often require a company to engage in international activities at an early stage, in order to reach a critical mass of users and set de-facto standards.

## 2.2.2 Internationalization

The internationalization of companies can primarily mean two things: either entering a new foreign market to generate revenue abroad, or opening company locations of different kinds (e.g. for distribution or production) in other countries. In a study on the internationalization of European small and medium-sized companies conducted by the European Commission, the authors have an even broader definition of internationalization. They include all meaningful business relationships with foreign partners: exports, imports, foreign direct investments, international outsourcing, and international technological cooperation (European Commission 2010).

To assess the internationalization of young software-based companies, it seems more appropriate to draw on the revenue abroad as a benchmark instead of all relationships with foreign partners. Consistent with many studies on international high-tech companies (Johnson 2004; Knight et al. 2000; McDougall 1989; Oviatt and McDougall 1997; Zahra et al. 2000), this research project employs a distributional definition. The specific share of international revenue above which a company is considered an "international venture", in contrast to a "domestic venture", depends on the author and the study, and varies between 5 and 20 %. Accordingly, a company's degree of internationalization was measured in this research project as the share of foreign revenue relative to a company's total revenue.

It was important that our study did not omit business models that have recently emerged, especially freemium- and advertising-based business models, which are based on offerings that initially are free of charge to the customer. Therefore, the case studies that we conducted also included companies that were not yet generating revenue abroad, but had been able to gain customers in other countries. In this context, foreign markets are all international markets that are outside the domestic market. A higher degree of internationalization thus corresponds to a larger share of foreign revenue respectively a larger share of foreign customers.

## 2.3  Current State of Research

To our knowledge, there has not been any research in which our research question has been explicitly addressed from a holistic perspective. At the same time, there is a plethora of research and a large number of studies that are highly focused on individual aspects. From the beginning of this study, we built upon these findings while conducting secondary research. As these findings are the starting point for deriving our results and recommendations, we will briefly summarize them in the following section.

### 2.3.1  Literature Review

There are a large number of theories in the literature on the internationalization of companies (e.g. Calvet 1981; Chetty and Hamilton 1993; Holtbrügge 2003; Kappich 1988; Miesenbock 1988). The theoretical foundations include the product lifecycle theory (Vernon 1966), the internationalization theory with a focus on market entry (Buckley and Casson 1976), the diamond model (Porter 1990), the eclectic paradigm (Dunning 1973, 1988, 2001), the Uppsala model (Johanson and Vahlne 1977, 1990), the network theory (Johanson and Mattsson 1988), and the theory of international new ventures, which is also referred to as "born globals" theory (Oviatt and McDougall 1997). These individual theoretical concepts deal with various forms of internationalization. While most concepts focus on specific aspects of internationalization, only a few claim to be universally valid on an abstract level.

Furthermore, there are a large number of studies on the growth and internationalization of companies. In the section that follows, we provide examples of studies that all highlight different reasons for successful growth and/or internationalization. These reasons can be broadly grouped into the factors strategy, financing, networks, and company size. All of these factors are also addressed, directly or indirectly, by our study.

#### 2.3.1.1  General Internationalization Strategies

Bell et al. (2004) point out that the internationalization strategy of knowledge-intensive companies needs to be analyzed separately. In contrast to traditional companies that react passively and are characterized by opportunistic "*ad-hoc*" internationalization, knowledge-intensive companies internationalize faster and follow a more structured internationalization strategy. Drivers of the growth of software companies are likely to be found in the strategic areas of innovation management, internal processes, and marketing, as well as in cooperation with other companies and research institutes, more freedom for employees in research and development, a higher investment in marketing, and an increased use of risk capital (Holl et al. 2006). The results by Ojala and Tyrväinen (2006, p. 79) further suggest a connection between a software company's mode of market entry and its product strategy: companies whose product strategy is based on cooperating closely with clients make use of their own sales representatives, while companies that sell standardized mass-market products to end users prefer cooperating with local partners. In addition, Stiehler et al. (2009) identify specialization, a systematic innovation process, and an early and active internationalization as promising business strategies to become internationally successful. On the other hand, especially large corporations benefit from a business strategy that includes outsourcing solutions (Kurbel and Nowakowski 2012).

### 2.3.1.2 Financing a Company's Internationalization

One study that was limited to an analysis of German companies found that the high cost of entering new markets is the main impediment for the German information and communication technology (ICT) sector to engage in international activities (Bertschek et al. 2011, p. 55). Another study by Metzger et al. (2010) emphasized the negative impact of a lack of debt financing, a lack of access to equity investors, and insufficient funding for young Germany high-tech companies. Jung et al. (2009, p. 28) conclude in their survey among 200 experts that a lack of financing is a "*substantial impediment to found a company*" for entrepreneurs in Germany.

The type of financing also seems to have an impact on the way a young company develops. Companies that are backed by venture capital more quickly and more frequently implement measures to increase the professionalization of the business, such as systematic human resource policies, stock option programs for employees, and even the employment of external managers (Hellmann and Puri 2002). Not implementing such measures of professionalization could increase the time that it takes to internationalize. Egeln and Müller (2012, p. 4) find that particularly those German companies in the ICT industry with the highest potential for growth suffer more often from a lack of shortage of capital. The large majority of companies with only a limited potential for growth, however, are less likely to suffer from a scarcity of capital, as their expansion plans are less ambitious.

### 2.3.1.3 Networks

Loane et al. (2004) identify a qualified management team with international experience and networks as the most important success factors for internationalization in a competitive environment that is quickly becoming more international.

International contacts and a talent for identifying international business opportunities are both seen as a critical driver of successful internationalization. Johnson (2004) emphasizes in this context the importance of the founders' visionary goals and ambitions.

One factor that Onetti et al. (2008) point out is the role of *"serial entrepreneurs"*. These serial entrepreneurs have already founded—with or without success—a company several times. They can therefore utilize their abilities that they have developed when it comes to internationalizing a company. This also relates to the role of established networks, which can influence the type and selection of a target market (Moen et al. 2004). These networks can be one key factor for small and medium-sized companies to gain an understanding of market selection, to choose the mode of market entry, and to achieve international efficiency (Amal and Freitag Filho 2010, p. 619). Hence, there is a connection between a company's national and international networks and its exports and market diversification. At the same time, Isenberg (2008) finds that it is especially the use of ethnic networks that has a positive impact on internationalization. Gabrielsson and Kirpalani (2004) show that multinational corporations acting as system integrators and distributors can be an option for companies to establish partnerships. Moreover, the Internet facilitates establishing networks as well as conducting marketing activities, which are both important factors for targeting unknown international markets. Sharma und Blomstermo (2003) find that *"born globals"*, which are companies whose strategy focuses on an international expansion early on, require knowledge on the mode of market entry before they actually enter the first market. They therefore make the decision about the mode of market entry based on their own and/or on their network's knowledge. They acquire additional knowledge through drawing on clients' and partners' networks.

### 2.3.1.4 Speed of Internationalization and Company Size

In a study on the internationalization and the export growth of small companies, Andersson et al. (2004) find that dynamic environments that are changing fast foster the internationalization of companies. Besides, the conditions underlying the speed of internationalization in early phases are different from the ones in later phases. A study on the internationalization of small software companies by Bell (1995) concludes that by following a larger partner organization or client, software companies can possibly overcome a stepwise entry into markets that are psychologically and geographically close. Autio et al. (2000) find that the earlier companies face international competition, the more quickly they grow internationally.

According to Coviello and Munro (1997, p. 379), it is especially young software companies whose internationalization follows a quick and stepwise approach. The speed of internationalization is accelerated or slowed down by the company's network. The analyzed companies had started to operate on an international scale within on average 3 years of their inception. Ruokonen et al. (2008) also find that small high-tech companies internationalize rather early, as long as attention is given to the following three elements: customer focus, focus on competitors, and coordination of the value creation network. These companies' entry into foreign markets was initiated by the globalization of markets on the one hand and the technological progress in information and communication technologies on the other. In order to make optimal use of these aspects and to internationalize early on, key factors are a company's innovation culture, as well as knowledge and skills regarding the internationalization (Knight and Cavusgil 2004). According to Knight and Cavusgil (2004, p. 137), traditional impediments to internationalization, such as a lack of experience or of resources, are of minor importance, as long as the strategy and culture have been directed at the company's internationalization at an early stage.

## 2.3.2  Research Gap

Our analysis of the literature shows that a lot of existing research has analyzed some important aspects also found in this research project. However, some relevant questions still remain unanswered. In spite of encompassing theoretical and empirical foundations, the existing approaches are, as we pointed out, fragmented (Kutschker and Schmid 2011, p. 380). Although some efforts have been made to integrate these focused approaches, their explanatory power suffers in part from a weak theoretical foundation, as the topic has often been approached phenomenologically (Jones et al. 2011).

In addition, there is a lack of scientific studies that emphasize the characteristics of small and medium-sized companies in the software-based industry and how their characteristics relate to internationalization approaches. A lot of older internationalization models still concentrate on traditional industrial companies (e.g. Amal and Freitag Filho 2010) or do not take a company's size into consideration (e.g. Johanson and Vahlne 1977). There have been efforts to account for small respectively young high-tech companies (Johnson 2004) and even to analyze the software industry in particular (e.g. Moen et al. 2004). However, many of these studies are based on a small number of case studies (e.g. Coviello and Munro 1997), which makes further research on internationalization strategies of software-based companies necessary.

Moreover, there is a lack of research conducting an international comparison of internationalization strategies (e.g. Coviello and Munro 1997; Ruokonen et al. 2008). Differences in the specific conditions of the home countries are frequently not taken into account. A study may look at one or two, sometimes very particular, countries (e.g. Gabrielson and Kirpalani 2004; Bertschek et al. 2011). This approach reduces the studies' generalizability and comparability. With this in mind, our study puts an emphasis on comparing ecosystems and best practices across seven countries, in order to develop specific recommendations for German software-based companies. The findings can then be potentially applied to other economic regions.

Besides, existing studies mainly deal with either impediments (Metzger et al. 2010) or success factors (e.g. Coviello and Munro 1997; Loane et al. 2004) and analyze the importance of individual factors without considering the context (e.g. Egeln and Müller 2012) and the interdependencies among the factors. This is the reason for the lack of an integrative approach that aggregates macro- and micro-economic perspectives and analyzes the special characteristics of software-based companies' internationalization strategies holistically.

Finally, there is a lack of studies that consider several dimensions and concisely derive actionable recommendations. From a methodological point of view, there are only few studies which combine quantitative research with qualitative observations in order to derive holistic results (e.g. Andersson et al. 2004; Jung et al. 2009).

Based on these considerations, this study focuses on the following research question: what are the critical factors for the growth and internationalization of small and medium-sized software-based companies, and through which managerial measures and through what kind of institutional environment can their growth and internationalization be fostered most efficiently? By addressing this research question, this study sets out to contribute to existing research in this area. The employed methodology draws on an integrative approach that tightly interlocks qualitative and quantitative research. The next chapter will explain in detail how this methodological approach was put into practice.

# Chapter 3
# Methodology

## 3.1 Project Structure

The research project was structured in two phases, which built upon each other and were linked with regards to their content. At the same time, the two phases differed in terms of their methodological approach and the specific research focus (cf. Table 3.1).

The first phase focused on identifying the factors that affect the internationalization of software-based companies in Germany. We set out to answer this question with a dedicated large-scale empirical study.

As a first step, we reviewed the existing scientific literature and conducted semi-structured expert interviews with around 75 entrepreneurs, investors, representatives of industry associations and incubators, professors, and experts from other areas.

Based on the results, we developed a questionnaire and conducted an online survey among the managers of German software-based companies (N = 1,064). The statistical analysis of the collected data allowed us to identify the relevant factors that have a significant impact on the internationalization of software-based companies in Germany.

Following this approach, we could determine both the drivers and the barriers of internationalization (cf. Chap. 4). The results of the first phase were first presented to an expert audience prior to the government's IT summit in Munich in December 2011.

The second phase built upon the findings of the previous phase and focused on the question of how small and medium-sized companies can internationalize successfully. Between January 2012 and June 2013, we conducted 55 cases studies—25 of which were in-depth—with German and foreign software-based companies that had internationalized successfully, in order to analyze their strategies regarding patterns and commonalities. Because of the segments in which expert interviews and literature research indicated the biggest potential for Germany to develop

© Springer International Publishing Switzerland 2015
A. Picot et al., *The Internationalization of German Software-based Companies*,
Progress in IS, DOI 10.1007/978-3-319-13548-9_3

**Table 3.1** Overview of the research project's two phases

|  | Project phase I | Project phase II |
|---|---|---|
| Time period | January 2011 until December 2011 | January 2012 until June 2013 |
| Research question | Which factors affect the internationalization of software-based companies in Germany? | How and through which strategies can small and medium-sized software-based companies in Germany internationalize successfully? |
| Methods | • Literature analysis<br>• 75 Expert interviews<br>• Quantitative online survey (N = 1,064) | • 55 National and international case studies, 25 of them in-depth<br>• 22 Expert interviews |
| Objective | Identification of the factors that affect the internationalization of software-based companies | Development of recommendations for companies and the public sector (i.e. the government as well as trade/industry associations) |

international market leaders, focus was put on the segments online games, web-centric services, as well as security and application software. We were able to identify several positive examples in Germany in these segments. Furthermore, interviews with 22 experts (mainly abroad) were conducted to complement the information gained in the case studies. The qualitative analysis of this comprehensive dataset made it possible to develop recommendations for different groups of stakeholders—both companies and the public sector—to point out the success factors for overcoming impediments to the internationalization of software-based companies. Preliminary results of the national case studies were first presented prior to the government's IT summit 2012 in Essen.

The results of both phases of this project are combined in this final report. They are structured and presented according to the focal areas capital, human resources, product strategy, scalability, growth strategy, cooperation partners, and ecosystem. Thereby, the relevant aspects of the internationalization of software-based companies can be highlighted and at the same time the findings can be presented in an integrated way. The two phases constitute a homogenous project. However, separating the two phases makes it easier to recognize the logical and chronological structure of the project. The following Sects. 3.2 and 3.3 will present both phases in more detail.

An overview of the research approach is shown in Fig. 3.1.

The research process began with a comprehensive literature review, which was followed by exploratory expert interviews and the quantitative and qualitative study. With the research project's progress, the information from the different methods was increasingly integrated. As an example, the results of the quantitative survey formed the basis for the guideline for the following case studies, in which open strategic questions were addressed. Accompanying literature research ensured that the status quo in the academic literature was considered, while expert interviews allowed us to continuously integrate the expertise of international experts and develop recommendations.

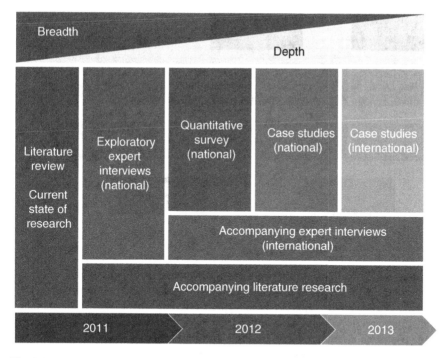

**Fig. 3.1** Focus and research methods in the course of the project

## 3.2 Details on Project Phase I

The first project phase focused on designing and conducting a large-scale quantitative survey. This survey was carried out with a web-based questionnaire among German software-based companies. Below, we will describe in detail the development of the web-based questionnaire, the selection of the companies, the actual online survey, and the data analysis. Figure 3.2 illustrates the most important elements of the research process in the first phase.

### 3.2.1 Development of the Online Questionnaire

The development of the measurement instruments was built upon two fundamental steps: on the one hand, we conducted an extensive literature analysis with a focus on both theoretical and empirical research, in order to review relevant research on the research topic, identify potential factors of influence, and build upon elements from questionnaires that had been used before.

On the other hand, we conducted a qualitative pilot study in the first half-year of 2011 consisting of 75 semi-structured expert interviews. As we set out to gain a

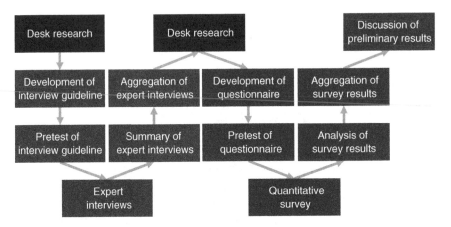

**Fig. 3.2** Research process of phase I—expert interviews and quantitative survey

perspective as broad and open as possible, we interviewed experts with different backgrounds (ICT companies, venture capital, universities, and policy-makers). The interview partners were selected based on an ongoing, theoretically sound comparative process (cf. Strauss and Corbin 1990). We paid special attention to strike a balance between different experts from the software industry, managers of different company sizes (small, medium, large), different industries and different stages of internationalization, as well as risk capital investors from different phases (seed, growth, later stage). After the first interview phase, additional interview partners were identified from the initial interview partners' networks. Thus, we could multiply the number of interview partners by drawing on an increasing number of networks (cf. Berg 2006; Bryman 2004).

Interviews were conducted in person on-site, except for cases where interview partners were not available for personal interviews; these interviews were conducted via telephone. To reduce social desirability bias, interview partners were assured anonymity. The interview was audio recorded and field notes were taken.

The guideline for the expert interviews was based on a further development of Michael Porter's theory of national competitive advantages, the so-called "diamond model" (Porter 1990). Our model combined micro- and macro-economic factors and allowed us to structure and analyze the drivers of growth and internationalization, both internal and external to a company.

We developed, to acknowledge Porter's conceptual model, the interview guideline shown in Fig. 3.3. The following factors, which we predicted to directly and indirectly impact a company's development, were part of the expert interviews and critically reflected: company-specific factors, industry, networks and contacts, a company's physical location, society, and government.

In order to make the interview guideline more specific and to reach a level of abstraction that is appropriate for practitioners, several topics based on the model were generated in the next step.

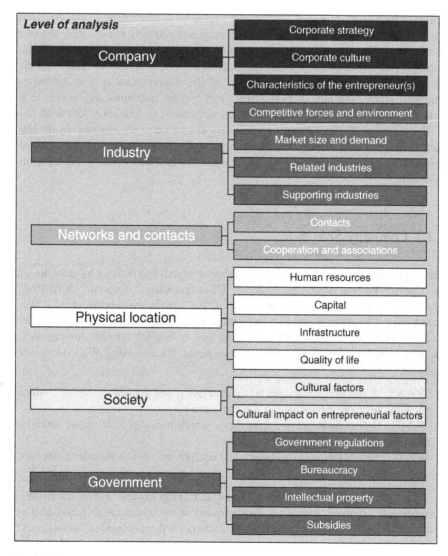

**Fig. 3.3** Structure of the expert interview guideline

The interview guideline was pretested and a final version was created. To account for interviewees' availability and area of expertise, the interviews were slightly adapted where necessary. As answers started to become increasingly repetitive between the 60th and 75th interview, we assumed that theoretical saturation had been reached. The guideline for the expert interviews can be found in the appendix.

The next step consisted of creating a collection of questions based on the findings from the expert interviews and extensive preparatory work. We avoided using hypothetical wording as well as subjective assessments and opinions wherever possible, in order to not create empirical artifacts but identify the true factors that influence the internationalization of software-based companies. Questions were therefore asked in a way as to assess the true state of a company and the environment in which it operates. Based on the requirements of a quantitative-empirical study, we almost exclusively used closed questions and scales (e.g. semantic differentials, 5-point Likert scales, selections, or free-entry text fields).

The software "SoSci Survey" respectively the server www.soscisurvey.de was used to convert and display the manuscript of the survey as a web-based questionnaire and to guarantee that scientific standards and data privacy requirements were met.

## 3.2.2 Company Selection

The total population of the companies to be surveyed was defined by drawing on the German Federal Statistical Office's "Classification of Economic Activities", which is grouped by commercial sectors ("WZ" from the German term "Wirtschaftszweig"). As our study's definition of software-based companies also includes web-centric application companies and embedded systems companies in addition to software companies in a narrow sense, the following WZ codes were included:

- WZ 62: "Computer programming, consultancy and related activities", which includes software companies in a narrow sense,
- WZ 63: "Information service activities", which includes web-centric services, and
- WZ 28: "Manufacture of machinery and equipment", which includes embedded systems.

To select relevant companies, we used the "Hoppenstedt Firmendatenbank" (Hoppenstedt company database). This allowed us to extract the required contact data and send a personalized e-mail to members of the companies' management team, inviting them to participate in the survey. One has to keep in mind that the Hoppenstedt company database does not include very small, i.e. "micro" (European Commission 2014), companies that employ less than 10 employees. As the research project's main focus was on small and medium-sized companies, this limitation can be neglected. Despite the high quality standard of the database, one cannot rule out duplicate, missing, or wrong records. As our survey builds upon the database, it cannot be considered completely representative. Because of the selection based on the WZ codes, not all companies could be considered part of the target population. Therefore, a "software matrix" (cf. Table 3.1) was developed, based on which companies could classify their business and indicate their main revenue sources.

### 3.2.3  Online Survey

After conducting a successful pre-test, we sent a personalized e-mail invitation, which contained a link to the survey, to the management of the selected companies. As an individual code was attached to each link, we ensured that the survey could only be taken by each participant once; however, it was possible to resume a suspended session at a later date. As each code was assigned to a company randomly and was saved separately, the respondents remained anonymous.

A total of 1,064 companies participated in the survey between September 19 and October 18, 2011 and completed the survey. The response rate of 11 %—accounting for invalid contact persons—can be considered a high value for an online survey. The retention rate, i.e. the share of companies participating and completing the survey, was 21 %, which can also be considered very good.

### 3.2.4  Analysis

In order to increase the quality of the dataset, cases with more than 25 % missing values, i.e. questions that were not answered, or cases where the survey was not completed and less than 80 % of the survey pages were answered, were excluded from further analysis. In addition, only cases where it took at least 500 s to complete the survey were included in the further analysis, as answering the survey too quickly raises serious doubts about the quality of the answers. The dataset was also checked manually for outliers.

The data cleansing resulted in 869 companies that were included in the further analysis. About 80 % of these companies generated a total revenue of less than 10 million Euros in their last fiscal year, and about 70 % of the companies employed less than 50 employees (cf. Table 3.2). The companies can be classified into the following segments (multiple selections were possible): embedded systems (33 %), system software (45 %), web-centric applications (47 %), and application software (63 %).

The dataset was analyzed using different statistical procedures (among them descriptive statistics and both binary and logistic regression). A linear regression was conducted to determine the factors which impact the internationalization of software-based companies. The dependent variable (degree of internationalization)

**Table 3.2** Distribution of company sizes in the sample

|                        | Revenue           | Share (%) | Employees | Share (%) |
| ---------------------- | ----------------- | --------- | --------- | --------- |
| Very small companies   | ≤2 million euros  | 43.6      | <10       | 18.0      |
| Small companies        | ≤10 million euros | 36.3      | <50       | 53.8      |
| Medium-sized companies | ≤50 million euros | 14.7      | <250      | 20.6      |
| Large companies        | >50 million euros | 5.4       | ≥250      | 7.6       |

was measured as the relative foreign revenue, in order to avoid biasing the results in regard to company size. Several independent variables from different areas (capital, human resources, product strategy, scalability, growth strategy, cooperation partners, and ecosystem), some of them aggregated to indices, were entered into the regression model; the variables had been identified earlier as potential influential factors. The results of the statistical analysis will be explained in detail in the individual sections in Chap. 4.

## 3.3  Details on Project Phase II

Case studies with national and international software-based companies that had already internationalized successfully were central to the project's second phase. The second phase built upon the findings of the project's first phase and expanded on them with qualitative empirical research methods, in order to derive specific recommendations for different stakeholders. The following sections provide details on the development of the interview guideline that was used in the case studies, the selection of the companies, the interviews themselves, as well as their analysis. Figure 3.4 shows this study's most important elements of the second phase of the research process and the integration of the findings into the final report.

### 3.3.1  Development of the Case Study Interview Guideline

An interview guideline was developed before the case studies were conducted. Before conducting each individual case study, background research was conducted on the respective company. Each case study consisted of (several) qualitative interviews with interview partners from different areas of the company (management, human resources, law, finance, product development, sales, and marketing). The interview guideline was structured to address each area of the company separately, thus ensuring that the interviews were conducted consistently and reliably across different interviewers. By structuring the content of the interviews, the interview guideline facilitated the subsequent analysis and reduced the risk of interviewees' getting stuck on one topic.

To determine how and through which strategies small and medium-sized software-based companies are able to internationalize successfully, the questions were open-ended. In using this approach, we did not restrict or possibly bias the interviewees' answers. It also allowed for exploring aspects that had not been considered before, thus maximizing the insight gained from each interview. At the same time, each question in the interview guideline contained important key words. These key words were marked as covered once an interviewee touched upon this subject. When key words were not addressed by the interviewee, they were brought up by the interviewer before proceeding to the next question.

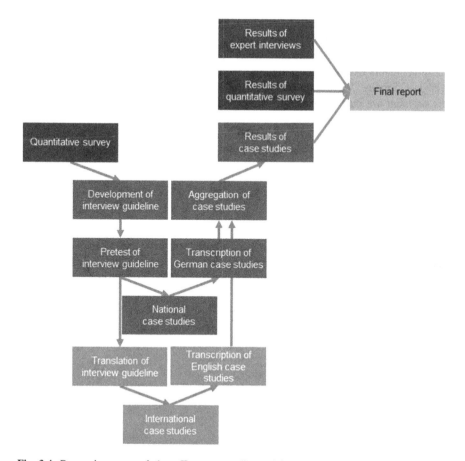

**Fig. 3.4** Research process of phase II—cases studies and final report

The content of the interview guideline was developed based on the expert interviews, the online survey, and the results of the first project phase. Special attention was given to the factors that had a significant impact on the internationalization of software-based companies. In contrast to the first phase of the project, which focused on the positive and negative factors influencing the internationalization of software-based companies, the case studies centered on the "why". The interview guideline can be found in the appendix.

## 3.3.2 Company Selection

To derive "recipes for success" and recommendations, the second project phase focused on German and foreign software-based companies that had already internationalized successfully. As the project's focus was not on company creation, rather on growth and internationalization, the main emphasis was on small and medium-sized companies. The companies that were selected derive from the following segments: games, Internet platforms, security and application software. These segments exhibit high potential and many positive examples can be found, even in Germany. We did not focus on embedded systems, as many companies in this segment are part of large corporations (e.g. Siemens or Bosch), which would have required a different research approach.

Especially in Germany, we analyzed companies from different regions, so that we did not bias our findings by concentrating on only one particular region.

### 3.3.3  Case Studies

Between January 2012 and June 2013, we conducted a total of 55 case studies in Germany and abroad (cf. Table 3.3 and Fig. 3.5). A case study could consist of up to seven interviews in a given company, depending on how many individuals were

**Table 3.3**  Number of case studies by countries/regions

| Country | Germany | USA | Great Britain | Russia | Israel | Scandinavia | Bulgaria |
|---------|---------|-----|---------------|--------|--------|-------------|----------|
| # | 21 | 21 | 4 | 4 | 2 | 2 | 1 |

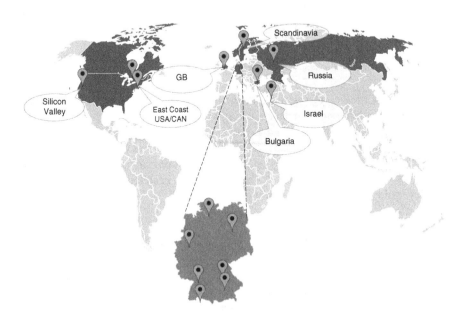

**Fig. 3.5**  Locations of the case study companies

responsible for the areas general management, human resources, legal, finance, product development, sales, and marketing. Therefore, the interviews were of different duration and included different parts of the interview guideline, depending on which areas the interviewee was knowledgeable about. On average, the total duration of the interviews was about three hours per company.

### 3.3.4  Analysis

The objective of the interview analysis was twofold: first, extracting from the case studies the company-specific recipes for success as "best practice". Second, identifying across all case studies those common patterns and similarities which foster a software-based company's successful internationalization. Therefore, both an individual-level analysis and an analysis to generalize our findings (cf. Lamnek 2010) was conducted.

The transcripts were first prepared qualitatively. Text passages that did not follow the interview guideline were grouped thematically. In addition, central passages were highlighted and commented on by the researcher. This approach resulted in a text which was reduced to the central themes. This text served as the basis to interpretatively extract the recipe for success of the interviewed company as best practice.

The analysis to generalize the findings was also based on transcripts that were qualitatively prepared by categorizing individual text passages and assigning them, according to the interview guideline, to different focal points. This resulted in a large number of structured pieces of content, which were analyzed by their frequency. The resulting patterns were examined critically and aggregated to a higher level of abstraction, in order to conclude universal success strategies and recommendation. A total of 25 cases studies were analyzed in depth following this approach. The findings from the case study phase, supplemented with the findings from the quantitative survey of the first phase, will be presented in the following chapter.

# Chapter 4
# Results

When identifying the determinants of software-based companies' successful internationalization, one can distinguish between factors external and internal to a company, as well as the interplay of these two domains. Regarding the external factors, we focused on the input factors capital and human resources. Regarding the internal factors, our main focus was on product strategy, scalability, and growth strategy. Regarding the determinants that deal with the interplay between the internal and the external domain, we concentrated on cooperation partners and the economic ecosystem.

As they represent input factors, external factors are described mainly on a (macro-)economic level. However, they also include strategic aspects, as dealing with these external determinants is a company-specific strategic decision. The internal factors, on the other side, are analyzed almost exclusively on a business level, as the focus is on a company's management. The determinants that deal with the interplay of internal and external factors bridge the internal and external domain and describe the different stakeholders' interaction within the ecosystem.

Similar to a biological ecosystem, an ecosystem can be understood as a balanced regional system of actors and environmental influences. The ecosystem has functional characteristics and certain regulatory mechanisms, which help it remain in a self-reinforcing balance (Iansiti and Levien 2004). An innovation ecosystem, such as the software industry in a certain region, can be perceived as a dense and multi-faceted network of private actors (entrepreneurs, venture capitalists) and public institutions (universities, public research centers), that strive to foster technology development and innovation. These actors jointly keep the ecosystem in a stable and self-sustaining balance (Jackson 2012).

This chapter is structured as follows: each section focuses on the study's central findings regarding a specific success factor. At the beginning of each section, we present the results of the statistical analysis from the first project phase. We then identify, based on the analysis of the qualitative case studies, patterns to explain and interpret the empirical results. Based on our analysis, we derive specific recommendations for companies and the public sector at the end of each section. To illustrate the sometimes rather abstract relationships, we additionally draw on

© Springer International Publishing Switzerland 2015
A. Picot et al., *The Internationalization of German Software-based Companies*,
Progress in IS, DOI 10.1007/978-3-319-13548-9_4

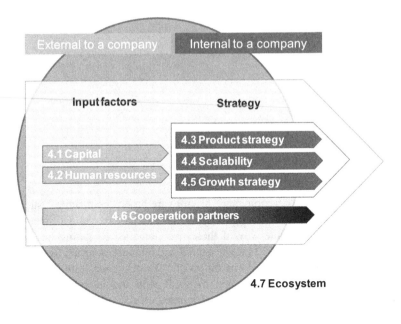

**Fig. 4.1** Overview of the following sections

specific examples from the qualitative case study phase to demonstrate best practices.

The sequence of the following sections is structured as depicted in Fig. 4.1, first focusing on the factors external to a company and then on the factors internal to a company.

## 4.1 Capital

The question of whether there might be a lack of capital within the German software industry is a contentious issue. This claim is backed by a considerably lower degree of internationalization of software-based companies from Germany compared to their global competitors (cf. Sect. 2.1). The findings from the quantitative study shed light on this issue.

### 4.1.1 Quantitative Results

The descriptive analysis shows a mixed picture regarding the German software-based companies' need for capital. 40.4 % of the surveyed companies considered a shortage of capital as an impediment to growth. On the other side, 44.0 % of the

Does a lack of capital pose an impediment to growth?

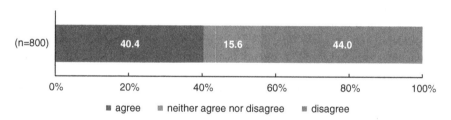

**Fig. 4.2** Descriptive statistics—lack of capital as growth impediment

companies did not regard a lack of financial resources as an obstacle for further expansion of their business. 15.6 % of the surveyed companies were neutral on this topic (cf. Fig. 4.2). While 40.4 % reported a shortage of capital, the regression analysis did not reveal a significant statistical impact on companies' degree of internationalization.

Based on the analysis, we conclude that companies that consider the lack of capital as an impediment to growth have not internationalized to a smaller degree than companies that do not report a lack of capital.

Thus, the statistical analysis does not substantiate empirically the correlation between the reported lack of capital by some software-based companies and their international expansion. The results of the qualitative case study analysis can shed light on why the lack of capital perceived by some of the surveyed companies is not the reason for their (in)ability to internationalize.

### 4.1.2 Qualitative Case Study Results—Germany

Managerial prudence and the pursuit of entrepreneurial independence dominate the growth and internationalization process in Germany (cf. "growth strategy"). For this reason, activities to expand a company are preferably financed with capital from "friends and family" along with bank loans. This funding structure profoundly impacts the behavior of software entrepreneurs in terms of the following three factors:

First, because German companies are financed internally instead of by risk capital, gaining positive cash flows is an essential part of their business strategy.

Second, the mode of financing also fosters a niche strategy: founders deliberately focus on entering individual market segments in which they can possibly gain market leadership.

Third, internal financing facilitates the focus on a sustainable growth strategy. This results in a successive but also occasionally unfocused approach when entering international markets.

On the one side, internal financing fits with German software entrepreneurs' desire to remain autonomous and enables them to make managerial decisions independently. On the other side, this entrepreneurial independence also means that management cannot draw on a network of investors and external shareholders with international contacts and experience.

In addition to providing financial means, shareholders can also provide access to know-how and contacts, thus facilitating a successful internationalization (cf. "cooperation partners"). Therefore, one crucial factor in this context is to what extent investors are interested in a company's internationalization. Companies which are financed exclusively by their founders often lack an external network of consultants and shareholders to support the management team, inject external know-how into the company, and at the same time systematically drive the company's growth process.

What is the reason that no significant relationship between a perceived lack of capital and the degree of internationalization could be found? The answer lies in the entrepreneurial independence, sometimes even isolation, combined with a focus on sustainable growth and a niche strategy: within their strategy of sustainable growth, the surveyed companies do not struggle to finance their cautious and successive market entry.

Acquiring substantial funding is not a prerequisite for entering new international markets, especially for companies operating in B2B markets with no substantial network effects. However, this does not hold true for ambitious and risky internationalization strategies where internal financing and financing through bank loans reach their limits.

The situation is somewhat different for the analyzed German software-based companies operating in B2C mass markets with substantial network effects. In these instances, a shortage of (risk) capital can be the reason for not internationalizing their business successfully. In markets in which the utility provided by a digital product or service increases with an increasing number of users (positive network effects, cf. "motivation"), international market leadership can only be gained by internationalizing quickly and extensively. The analyzed companies' financial inflexibility impedes their aspired, and necessary, swift international expansion.

Moreover, the case study findings indicate that there seems to be a lack of domestic companies (in the ICT sector) that act as a strategic consolidator. Considering that many international software-based companies have reached their dominating position through strategic acquisitions, this aspect might be a critical factor as to why Germany is lacking a "mega player" with international visibility. Without a doubt, such a consolidation strategy would require a huge amount of capital.

Rather than giving a generic answer, as has been frequently done, regarding the importance of capital for the internationalization of German software-based companies, one must instead consider the study's findings and address this topic from a much more differentiated point of view.

### 4.1.3 Qualitative Case Study Results—International

In contrast to the market in Germany, the US market is characterized by a high availability of risk capital. The analyzed companies in the USA, mainly in Silicon Valley, had access to several times the growth capital (B and C round financing) that is commonly available in Germany. Financial sums that are invested in the so-called "C round" after 2–5 years as the foundation for internationalization are not available in Germany as growth capital; these sums would actually be deemed rather high for an exit, i.e. selling the whole start-up.

Due to the large amounts of capital available, the analyzed companies in Silicon Valley, which were always backed by venture capital, can afford quick and aggressive growth strategies. Generating revenue and a positive cash flow is not the focus of the case study companies that we interviewed. The objective is a profitable exit, selling the company to an acquiring company, after an average of 5 years.

Apart from the mere quantitative availability of capital in Silicon Valley, it is primarily the capital's qualitative dimension that matters to the companies analyzed in our study.

The board of directors of the respective companies is characterized by a variety of shareholders who are authorized to make decisions. As these shareholders do not receive the option to gain their initial investment back prior to a successful exit or taking the company public, all of them are incentivized to drive the company's quick growth. Based on how profits are distributed among scalable software-based companies and the experience that one out of ten companies in a portfolio is sufficient to make up for the other companies' losses and still multiply the capital invested, the slogan is "all-or-nothing".

In this system, the shareholders serve both as the drivers of the companies' growth as well as their advisors. The team of founders thus gains access to a wide network of experienced experts and industry specialists (cf. "cooperation partners" and "ecosystem").

The only money that will exist in twenty years is smart money.
(Business angel, Silicon Valley)

As they face different financial conditions, companies from Silicon Valley follow a different growth process than companies from Germany. The focus of managerial decisions lies on the fast growth of the companies. The decision of whether to grow nationally or internationally comes second. The shareholders usually jointly agree to enter those markets in which they expect the largest potential for success. At this point, one can add that the American market in itself offers enormous potential. If you add India and Great Britain as English-speaking markets and also individuals in other countries who speak English as a foreign language, one can imagine why the US market is so attractive for investors of risk capital.

Because of high profits gained by selling companies as well as many founders' aspiration to pass on their knowledge and know-how even after the sales, a thriving business angel scene has been established in Silicon Valley. Crowdfunding and crowdinvesting are two major trends within this scene.

The interplay of investors, experienced serial entrepreneurs, and highly skilled founders that all operate in the context of a financing system that fosters aggressive growth, along with large sums of available capital, can explain, to some extent, how the companies that we analyzed in Silicon Valley succeeded in growing quickly (cf. "ecosystem"). This form of risk capital financing particularly enables Internet companies in the B2C segments to grow quickly and aggressively. It is however surprising that even B2B companies from Silicon Valley act more aggressively and grow more quickly than similar companies in Germany. From the beginning, these companies focus on offering standardized products, employ standardized processes, and rigorously apply platform thinking. They are therefore in contrast with German B2B companies, many of which adapt their products to their clients' requirements.

We discovered a pattern of expansion similar to the one in Silicon Valley when analyzing companies from Israel. Using tightly woven personal networks within Silicon Valley and leveraging the access to American investors, the analyzed case study companies succeeded in entering the US market quickly (cf. "growth strategy"). The main factor for successful internationalization can be found in their access to local contact persons and business networks provided by their investors, in addition to their access to capital.

The companies that we studied outside Silicon Valley based in Great Britain, Scandinavia, and Eastern Europe show similar characteristics to German companies regarding their funding structure. Some companies from Moscow successfully followed a strategy of officially transferring their headquarters to the US west coast in order to make themselves appear to be American companies in the eyes of investors and clients.

### 4.1.4  Recommendations

Based on the study's findings, we can derive the following recommendation for companies and the public sector:

- Companies

  - German software-based companies that operate in markets with strong network effects and that are dependent on quick growth need to approach international investors. By being open for investments from international venture capitalists, companies can secure the funding for rapid internationalization. In addition, German companies can benefit from the international venture capitalists' networks. These networks can facilitate the company's access to customers, employees, and cooperation partners in the international target markets.

- Companies that plan to keep their independence and concentrate on sustainable growth can benefit from a focused niche strategy. Following this strategy, a company builds up deep knowledge in a selected market niche and subsequently uses this market niche as a stepping stone to gain international clients. This approach is beneficial, as the company is less likely to face international and financially strong competitors in this niche. Even without being backed by substantial amounts of capital, companies following this strategy can gradually enter new markets. By becoming internationally recognized experts, these companies can advance their internationalization self-confidently. Once a market niche has been occupied in the domestic and international market, the companies can expand into neighboring niches.

- Public Sector

  - To counteract the lack of capital, we recommend stimulating venture capital as a means of financing instead of direct market interventions by the public sector. Possible measures include a special asset class based on the French model, a change in the write-offs of losses, and national umbrella funds.
  - Specific measures to attract international growth capital should be developed in cooperation with international experts in the form of a venture capital panel. The objective should be to make financing German companies with venture capital as attractive as in other international markets.
  - A regulatory and legal framework should be created to make alternative capital markets, such as crowdfunding and secondary markets for company shares, more attractive. Legal certainty and consumer protection have to be guaranteed to gain and keep private investors' trust.

---

**Best Practice: "Reaching the Exit Together"**

100 % organic food. A young employee in her mid-twenties boastfully points out the quality of the food. As with all new trends, the Californian trend regarding a healthy lifestyle is being celebrated to the extreme.

In the software company from Palo Alto, the employees make themselves comfortable in the community kitchen. Most of them are in their twenties and mid-thirties and are dressed casually in colorful shorts and polo shirts. As in most companies in Silicon Valley, there is a large lunch buffet and a big American fridge from which everyone is invited to help themselves at no charge. The atmosphere is closer to a student dorm than to a multi-million dollar company.

"Think big!" is the employees' slogan, like everywhere else in Silicon Valley. Their goal is nothing less than getting the product, an app for mobile phones, among the top 10 of the most popular applications. The lively founder is excited: "Together we all drive the company's growth."

The smart founder from India with the tousled hair and a big smile on his round face moved to Silicon Valley several years ago to make his dream of his own company come true. Two years ago he founded the Internet company together with two graduates from Stanford University. He met them in one of the many bars that serve as a meeting point for the whole industry.

It is typical for so many companies in this region: a team consisting of two to three tech- and marketing-savvy founders, financed with risk capital, advised by a coach who is also invested in the company. Being financed with risk capital is a success factor, especially for companies that operate in markets with strong network effects and therefore need to grow quickly.

On the one hand, there is enough capital available for the management that they do not have to worry about sustainable profit. Instead, they can focus on growing quickly and gaining new customers. **The only thing that counts are the number of users**, along with the technology, for which large local corporations are willing to pay a considerable amount of money. On the other hand, this is the reason why there is a diverse team of **venture capitalists with industry experience on the board of directors**. They actively support the team of founders with their know-how.

**Everyone is working towards a common goal, the so-called "exit"**—taking the company public or selling it to a large corporation with the highest profit possible. Only then, the time and money invested by everyone involved will pay off—possibly multiple times the investment. The whole process of expanding the company, from founding the company until **selling it after a maximum of 5 years**, follows a structured plan. It does not matter whether the company operates internationally right from the beginning or initially focuses on the large domestic US market, which in itself offers a huge potential for growth. The only thing that matters is fast expansion.

**The financial incentives for the leadership team** and their prospects of a high price when selling the company will encourage everyone involved to provide their expertise to the company and drive its growth.

Capitalism "Made in America"—while enjoying ecologically correct hors d'oeuvres.

## 4.2 Human Resources

In addition to analyzing the access to capital, the study also focused on the importance of human resources. Recruiting highly qualified experts was crucial to the companies that we examined.

### 4.2.1  Quantitative Results

The descriptive analysis of the survey shows that the companies struggle to hire qualified employees. While 50.0 % of the companies that we surveyed agreed that they generally have difficulties filling all their open positions, only 23.4 % did not agree, and 26.6 % neither agreed nor disagreed.

To analyze the impact of the rather generic term "shortage of skilled staff" on German software-based companies on a more detailed level, we broke down the question into different categories. 56.3 % of the companies had problems recruiting employees that asked for an acceptable wage; only 14.3 % disagreed in regards to this statement, 29.4 % neither agreed nor disagreed. The findings were similar when analyzing the question of whether there are problems hiring staff that need little training on the job in order to become productive. 66.9 % agreed on this statement, while only 14.3 % did not agree and 18.8 % neither agreed nor disagreed.

Hiring employees with industry expertise and technical know-how posed the largest challenge. 75.3 % of the companies agreed that they struggle hiring employees with industry experience; only 9.8 % did not agree and 14.9 % neither agreed nor disagreed. The results were even more clear-cut regarding the recruitment of employees with specific technical know-how. 80.8 % of the surveyed companies had difficulties hiring technical experts; only 10.9 % did not face such problems, and 8.4 % were undecided on this aspect (cf. Fig. 4.3).

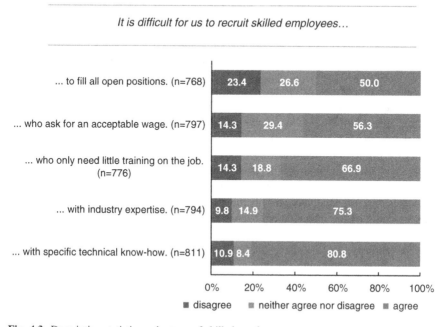

**Fig. 4.3** Descriptive statistics—shortage of skilled employees

Thus, the descriptive analysis of the quantitative survey shows that a shortage of skilled staff exists especially for experienced technical specialists.

Despite the reported shortage of skilled employees, the results of the regression analysis turned out to be non-significant. Thus, we could not determine a causal connection between the reported shortage of skilled employees and the software-based companies' degree of internationalization.

In an attempt to explain the reason why the supposed crucial problem for German companies, when it comes to recruiting suitable experts, does not cause their comparatively low degree of internationalization, the topic of "human resources" was made one of the central themes of the qualitative case study research.

## 4.2.2  Qualitative Case Study Results—Germany

The analysis of the German case studies shows that no general shortage of skilled employees exists. While we did not find a severe shortage of university graduates and general managers, the majority of the companies that we analyzed face the challenge of hiring highly qualified and experienced IT specialists. In some cases, software experts have to be intensively wooed by the companies to adequately fill open positions. Therefore, we find that software-based companies, to some extent, are confronted with the challenge of recruiting suitable experts.

Especially B2B companies that are financed by the founders themselves possess only limited financial resources to pay for the relatively high cost of experienced IT specialists (cf. "capital"). The high cost of recruiting suitable experts with many years of experience can explain why mainly small and medium-sized companies with relatively little financial resources complain about the shortage of skilled staff.

The situation for B2C companies that are financed by risk capital and that often need to grow quickly is somewhat different. A rapid expansion requires them to hire international employees in the areas of technology, sales, and marketing on short notice. Therefore, bigger cities in Germany become more attractive as a company location, as they offer access to a larger pool of suitable candidates (cf. "ecosystem").

At the same time, all analyzed companies react to the increasing competition for experts by improving their working conditions. This includes creating flat hierarchies as well as strengthening individual employees' responsibilities. We found that software-based companies in Germany expand employees' freedom to make their own decisions, especially regarding the positions of software developers. These steps are taken by the management to motivate their employees and simultaneously increase the companies' attractiveness as an employer for potential new hires.

In the majority of companies, suitable experts were primarily recruited within the management's personal network. Social networks serve as an important complement to job fairs and official job postings (cf. "ecosystem" and "cooperation partners").

In addition, some of the analyzed companies tried to selectively outsource activities that do not belong to their core business to companies abroad. By outsourcing simple programming tasks to a foreign company, a majority of the analyzed companies could circumvent the problems of hiring suitable employees in Germany. At the same time, the companies benefitted from these measures by achieving cost savings and gaining organizational flexibility.

### 4.2.3 Qualitative Case Study Results—International

The analyzed companies in Silicon Valley benefit from a large local supply of internationally experienced and highly specialized experts. The local companies can draw on a large pool of qualified employees, both regarding technical specialists and internationally acclaimed marketing and sales experts.

However, the analysis of the case studies shows that the companies have to face a substantial and well above average competition for these experts. The high demand for experts in Silicon Valley also causes potential employees to ask for high wages. Basically, the companies took three measures to attract qualified employees:

First, companies try to accommodate IT specialists' salary demands; because of the relatively high availability of capital, the required financial resource can usually be provided (cf. "capital").

Second, similar to German companies, companies in Silicon Valley try to establish an employee-friendly company culture. This includes flat hierarchies and strengthening individual employees' responsibilities.

Third, companies aggressively try to lure away specialists from their competitors, which makes management's local networks all that more important. Suitable candidates are often approached through personal contacts and informal networking events (cf. "ecosystem").

> The first guy is the most important hire. My philosophy was always hire the best guy from your local competitor if there's one. Just get him! Pay more. Get him!
> (Founder of a mobile advertising company, Silicon Valley)

At the same time, the high turnover of employees changing jobs between companies in the San Francisco Bay Area fosters the transfer of know-how across company boundaries. This high frequency, compared to German standards, of changing employers speeds up the diffusion of industry knowledge and diversifies business networks.

> So I think technical people are attracted by the brand and the technology roadmap. (…)
> Sales people are attracted by money. Yeah. Money and fringe benefits. And cool factor.
> (Founder of a mobile advertising company, Silicon Valley)

We could not detect any problems to recruit suitable employees in the analyzed Israeli companies. The companies indicated that they benefit from a large local pool of skilled experts. The Israeli armed forces play a crucial role in this context. The interviewed entrepreneurs emphasized that the technical training and assuming supervisorial responsibility, as well as gaining independence at an early stage during the military service are a locational advantage of Israel.

In contrast, Scandinavia and Eastern Europe presented a similar picture to Germany regarding the recruitment of new employees.

### 4.2.4 Recommendations

Based on the study's findings, we derive the following recommendations for companies and the public sector:

- Companies
  - In the short term, companies should increasingly leverage the potential of external service providers, offshoring, interns, and freelancers to outsource activities that do not belong to the companies' core competencies. This enables companies to reduce HR expenses on the one hand and allows them to act more flexibly on the other. Especially the risks of quick growth can be better balanced through this strategy.
  - In the medium term, companies should invest in a sustainable HR marketing strategy. This includes investments in recruiting events, cooperating with universities, and strengthening one's employer brand.
  - Companies should try to purposefully recruit the best specialists working for their competitors abroad.
  - To compete for the brightest minds on an international level, companies should raise a sufficient amount of capital to provide financial incentives early on.
  - In the long term, both monetary and non-financial incentives, such as flat hierarchies, are important levers to counter a shortage of skills and retain employees.
  - Companies should consider the transition to English as the company language as a measure to attract qualified experts internationally. English as the company language also facilitates the entry into global markets and opens other strategic options. However, one has to consider the cost of necessary training in order to compensate for a loss of linguistic precision and social interaction.
- Public Sector
  - In order to expand the pool of technical specialists in the forthcoming years, we recommend that the public sector offers a larger number of consecutive and specialized MINT (mathematics, information, natural sciences, and technology) programs of study.
  - By taking additional courses, business students can acquire technological knowledge that can be crucial for a company's product strategy, while software engineers can acquire leadership and marketing skills. Such

additional qualifications can be effectively taught through an "honors degree" in the area of technology management. Such a degree would accompany and supplement the regular (under)graduate courses in a suitable way. The Center for Digital Technology and Management (CDTM) offers such a format across universities and provides technical and business students with a cultural understanding of each other.

– The "Bologna Process" caused many universities of applied sciences to no longer require mandatory internships. Therefore, many software-based companies lost an important recruiting channel. To counteract this effect, vocational training, especially in the IT industry, should become increasingly more modularized. By being offered project modules with different software-based companies, apprentices can gain a maximum variety of industry experience and develop enthusiasm for entrepreneurship at the same time.

– Especially young start-up Internet companies that operate in markets with strong network effects and are therefore dependent on fast growth could benefit from flexible employment models and pursue a faster growth strategy.

---

**Best Practice: "Best Place to Work"**

This company is a little different. Despite the purism that the building emanates, everything seems to be more colorful. Standing on one of the upper floors with a view on the Alps, one has to feel cheerful—not only because of the Chinook that makes the mountains look so close to Munich, but also because of how friendly people interact, which has become one of the company's hallmarks. The company is active in the B2C segment of sales platforms and known far beyond Germany's boundaries.

The company has grown quickly in the last years. The first office quickly became too crowded for the many new employees. Is a shortage of skilled employees a problem for the company? "**There are a lot of applications from individuals with a business background**," tells the top executive. However, it is **difficult to hire specialists with IT know-how**.

The management decided to introduce new **organizational innovations** in order to be more attractive than the competitors. Instead of rigid hierarchies, employees' autonomy was strengthened. **In particular, IT developers are granted more freedom in order to realize their own ideas and work more independently**. At the same time, most teams are self-organizing. Colorful pieces of paper are stuck to **pin boards** in the hallways, where every team writes down the most important to-dos for the week and attends to them in cooperation with the other teams.

Management only interferes to coordinate and limits itself to monitor cooperation. This fosters a high degree of **personal responsibility** and contributes to employees' intrinsic motivation. This encourages particularly individuals with a university education, who are used to working independently and only unwillingly succumb to rigid hierarchies, to consider applying for a positing with this Munich company. **The relative scarcity of IT specialists is actively countered by introducing organizational innovations, which benefit all employees in the end**.

The teams are encouraged to **develop passion, make use of their freedom, and act creatively**. After all, a pleasant work environment which fosters individual freedom guarantees long-term stability. Flat hierarchies and innovative ideas regarding project work and team organization make the company stand out as an attractive employer for young and active employees. **By creating teams with interdisciplinary team members, the company can act more flexibly and more openly than other companies**.

"We offer the employees coachings, trainings, and conferences to further develop their skills," tells the executive and gleefully glances at the peaks of the Karwendel mountains, which can be seen through the window. The company is also setting out for new heights. By following new paths in self-organization and through finding creative ways to motivate employees, the company from Munich will continue to conquer many regions far beyond the Alps.

## 4.3  Product Strategy

After examining capital and human resources, which are factors mainly external to a company, the following three sections will deal exclusively with internal strategic decisions.

### 4.3.1  Quantitative Results

53.2 % of the surveyed companies answered the question regarding their focus in product development with "technical perfection". This contrasts with only 12.4 % of the companies focusing on "appealing marketing". 34.4 % of the surveyed companies indicated to focus both on technical perfection and on appealing marketing (Fig. 4.4).

However, German software-based companies' technical orientation is not disadvantageous. The regression analysis shows that neither a focus on technical perfection nor a focus on appealing marketing has a significant impact on companies' degree of internationalization.

Thus, companies focusing on technical perfection are not less international than marketing-oriented companies.

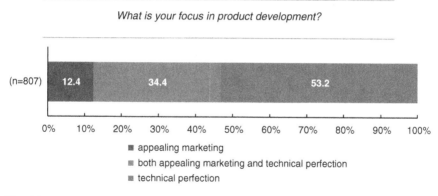

**Fig. 4.4** Descriptive statistics—focus of product development

The results of the case study analysis are helpful in interpreting these findings. The qualitative findings of our study allow us to draw insightful conclusions on the importance of marketing investments for software-based companies.

## 4.3.2 Qualitative Case Study Results—Germany

The qualitative analysis shows that there is a big difference between B2B and B2C companies regarding the impact of marketing.

The analyzed companies in the B2B segment focus mainly on the technical perfection of their products and solutions; therefore, the analysis of the case studies further validated the quantitative results. Moreover, the majority of the B2B companies consider marketing investments as investments of minor importance. Instead, they place the real focus of their company strategy on optimizing the quality of their products. This focus on the long-term development of their technology and a continuous technical optimization of their products is plausible given that software-based B2B companies are usually highly specialized.

In addition to analyzing the importance of marketing activities for a company's product strategy, we also examined whether the products were targeted at customers in international markets. We found that the analyzed B2B companies target their products primarily at customers in the German market.

The majority of the analyzed companies did not consider the companies' later internationalization at an early stage when developing their products (cf. "growth strategy"). The failure to do so results in many difficulties when handling the increasingly complex software once the company internationalizes.

In contrast, we noticed a stronger focus on marketing and usability among the analyzed B2C companies. Marketing and PR investment are not only seen as sales support, rather as an integral part of the business model. The objective is both to increase consumer awareness of the company and to create an appealing product for the end user.

Hence, the findings show that whether a company has a strong product focus depends to a great extent on the market segments and the business models of a company and cannot be generalized across the entire software industry.

### 4.3.3  Qualitative Case Study Results—International

In contrast to the German software-based companies, the companies from Silicon Valley are strongly PR- and marketing-oriented, regardless whether they operate in the business or end user segment.

Investments in increasing the level of awareness among clients as well as building a brand and a good reputation are also an important part of the B2B companies' product strategy. With these measures, companies aim to gain the attention and trust of potential new clients, employees, and cooperation partners (cf. "human resources", "cooperation partners", and "growth strategy").

In addition to building a strong technology, companies focus their product strategy on high usability and intuitive user interfaces. Simultaneously, a "good enough" approach dominates: rather than pre-planning products until technical perfection is reached, companies develop products jointly with their clients in an iterative fashion.

We also found that companies in Silicon Valley adopt new technologies very quickly. High employee turnover and cooperation among local companies and research institutes foster the transfer of know-how and the diffusion of knowledge, which are directly integrated in the product development process (cf. "cooperation partners" and "ecosystem").

Targeting products at customers in English-speaking markets simplifies the global expansion considerably and facilitates the setting of international standards. It has to be emphasized that the internationalization of products is built into the business strategy early on (cf. "growth strategy").

Early internationalization was also a part of the business strategy of the Israeli and Scandinavian companies that we analyzed. Because of their small domestic markets, they take a cross-border expansion of their business activities for granted and plan accordingly early on. Consequently, Israeli and Scandinavian companies are characterized by developing international and English-version products.

Half of the Russian companies that we studied were founded in Russia and stayed inside the Russian speaking market for a long period of time. This allowed them to gain as much of the growing market as possible, therefore benefitting from their proven products and business models. The other half of the analyzed companies, among them exceptionally successful examples, developed an English version early on, which enabled them to win customers around the world. After the first pilot customers in the USA showed interest in the software, these companies set up an office in the USA even before they had received growth capital. After being financed, they staffed this office with an experienced American manager and started a sales campaign. According to the founders, (Russian) companies should be internationalized in a way that de-emphasizes the Russian background, unless the company operates in technologically driven segments such as security or database software, in which the label "Software Made in Russia" stands for sophisticated "hacker" technology.

### 4.3.4 Recommendations

We derive several recommendations for companies and the public sector to optimize the product strategy:

- Companies

  - As international software can be easily compared, it is becoming increasingly more important to develop software as marketing and design statement that speaks for itself. To increase their chances in the global market, German companies with a strong technical focus that aspire to internationalize should make usability and design a high priority.
  - The products should be tailored to the global market early on. Based on the experience of Israeli and Scandinavian companies, it is advisable to develop an English language version. The software should allow the developers to change and add languages flexibly.
  - B2C companies, in particular, should consider entering international markets at an early stage and continuously improve their products in close cooperation with their customers (e.g. through releasing additional updates).
  - Marketing and PR play a key role in making a lasting contribution to the success of software-based companies. Therefore, companies need to understand PR and marketing measures as a central piece of their value chain. Investments in this area can be a substantial lever, even for technically oriented companies, to acquire new clients and international cooperation partners. To gain customers' trust, it might be necessary to invest in a US office.

- Public Sector

  - On a political level, interdisciplinary cooperation among university students should be fostered. Thus, an understanding of the relevance of PR and marketing activities for successful internationalization can be created, even to technically oriented founders. The possible integration of technology, business, and design into joint programs of study is just one example. As an alternative and to keep the high quality of regular (under)graduate courses, accompanying interdisciplinary qualifications could be offered.
  - German policies have primarily supported technology transfer from elite research institutions. Business model innovation and marketing, in contrast, have not been given much consideration. Simple technologies with innovative marketing-based business models have rarely been supported.

Competitions and programs that target new marketing-based ventures and support their international expansion would help to highlight the importance of marketing in business plans at an early stage.

### Best Practice: "Welcome to the Dynamic North"

Looking at Oslo from the Oslofjord, the snow-covered capital of Norway sparkles brightly on this sunny March morning. The ferry slowly approaches the dock. The ferry's determined and relaxed motions are similar to the atmosphere of this hidden Nordic champion. The company's headquarters, which is made of glass, is located on a small hill only a few meters from the harbor.

**Because the size of the Norwegian market is limited, the company was forced to aim at international markets early on**. This strategy benefits from both the digital means of distributing the products and the very international atmosphere in Oslo. "We have a lot of immigrated employees, some of which will go on to open new company offices in their home countries," tells the manager.

The down-to-earth Norwegian manages the company like a **well-organized family**. Similarly, the open-plan office is designed to be integrative and spacious. The employees meet in the cozy cafeteria, which is centrally located, or at the coffee station. A visitor does not find any bureaucratic structures. Instead, employees enjoy the freedom to make an impact. **The egalitarian structure of the Norwegian society is also reflected in the company's structure**.

The company is specialized in the development of mobile browsers which are distributed online. While other companies often spend a lot of financial resources on advertising and marketing to sell their products, this company attempts to **purposefully make use of viral marketing effects on the Internet**. In this context, the **relative market share in a target market is more important than the sum of all global users**. Through the local concentration of business activities, both business networks with customers and cooperation partners can be more easily established in the local markets, and viral marketing effects can be leveraged through personal recommendations.

By **cooperating specifically with international partner companies**, the company additionally benefits from their **cooperation partners' customer relations** and also the resulting distribution channels. "Through this approach, international awareness among potential customers can be increased quickly and inexpensively, and our product can be distributed more effectively," explains the manager.

**Viral marketing effects, cooperation, and a local focus**. The company from Northern Europe has come far by following this strategy. With a view on the capital from this postmodern office building, there is no doubt that the business years ahead will continue to be sunny.

## 4.4  Scalability

The scalability of the developed software is a requirement for fast internationalization. By analyzing the scalability, in addition to the product strategy, one can draw insightful conclusions regarding German software-based companies' growth problems.

### *4.4.1  Quantitative Results*

39.1 % of the surveyed German software-based companies replied that they only develop for the international market. 36.8 % responded that their products are only developed for the German market. 24.1 % of the companies develop products both for the German and the international market (Fig. 4.5).

The regression analysis shows a significant impact on the degree of internationalization. The analysis indicates that companies that target their products at customers in international markets early on also exhibit a higher degree of internationalization.

Whether or not German software can be easily internationalized is closely related to the question as to what extent products and business models allow for quick expansion across a country's borders. We were able to explore this topic in more depth with the qualitative case studies.

### *4.4.2  Qualitative Case Study Results—Germany*

The results from analyzing the German case studies reveal two contrasting findings. While some of the software-based companies face difficulties regarding their

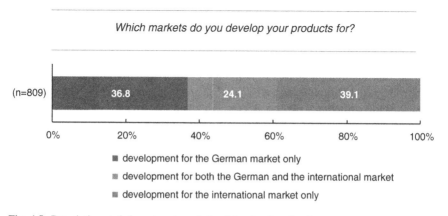

**Fig. 4.5**  Descriptive statistics—target markets of the developed software

products' scalability, other companies are very successful at implementing a business strategy that can be applied to new markets.

In particular, setting up local subsidiaries and staff-intensive services that are offered locally turn out tobe expensive and difficult to manage. Therefore, B2B companies that offer client-specific solutions and local customer support cannot easily scale up their business. Scalability would, at the same time, require large financial resources, which the surveyed companies struggle to raise (cf. "capital").

> We're a start-up, right. So we can't really have offices in every country. Our approach is to go with a partner first. Even in the UK, we have an office now. But initially we went with a partner.
> (Founder of a cloud security company, Silicon Valley)

In this context, the opportunistic acquisition of international customers is another factor that counteracts a planned and long-term standardization of the internationalization process (cf. "growth strategy"). Ad-hoc analysis of the target markets and rather short-term planning when entering new markets both tend to hamper quick international expansion.

In addition, the European market with its variety of languages, business cultures, and mentalities substantially complicates developing local product versions in comparison to the North American market. The majority of the analyzed companies therefore have difficulties in standardizing the internationalization process within Europe. It is for this reason that economies of scale can be achieved only, if at all, with great difficulty when attempting to take business models abroad.

> This is probably one of the key things, if you start in a small market and you start in a language other than English only, then it's probably tough. Right here it's a huge market. You can scale quickly.
> (CEO of a picture-sharing service, Silicon Valley)

In contrast, some software-based companies master these challenges very well. These companies internationalize without setting up staff-intensive and expensive local subsidiaries. Instead, the companies rely heavily on cooperation with local sales partners or domestic cooperation partners and use their infrastructure to gain access to local know-how. This approach enables the analyzed companies to keep a lean organizational structure while reaping economies of scale in the growth process (cf. "cooperation partners" and "ecosystem").

This context also shows the increasing importance of outsourcing staff-intensive services. Outsourcing of non-core activities enables the analyzed companies to become increasingly flexible and also fosters a quick expansion across geographical borders. Some companies rely on freelancers when employing companies abroad. However, this approach makes it necessary to monitor the quality and limit outsourcing to a small number of activities (cf. "growth strategy").

Besides, some of the software-based companies could increase the scalability by developing platform solutions. The software platforms add to the companies' flexibility and also facilitate location-independent customer service and virtual team work. By combining universal platform components with individual solutions, the companies manage to successfully balance scalability and individualized customer support (cf. "growth strategy"). Companies that apply these innovations also successfully reduce their staff requirements (cf. "human resources").

We also found that some companies are increasingly relying on English-language products in order to adapt more easily to global markets. At the same time, companies tailor their business models to a possible international expansion early on and thoroughly analyze the target markets in advance.

These findings strengthen the results from the quantitative study: those companies that plan for a global expansion early on and rationally plan and implement the process of internationalization successfully penetrate international markets.

### 4.4.3 Qualitative Case Study Results—International

The analyzed companies from Silicon Valley deliberately plan for early expansion and achieving economies of scale right from the start (cf. "growth strategy"). This includes, among other things, the implementation of platform solutions in connection with cloud computing, thus enabling cross-country and location-independent cooperation.

As another positive effect, platform solutions reduce the need to set up expensive international branch offices. Setting up branch offices is only considered once large local markets have been successfully entered and have shown enough potential. Apart from that, companies heavily rely on cooperation with sales and cooperation partners, whose business networks and sales infrastructure can be used to reach out and provide service to international clients (cf. "cooperation partners").

The companies, at the same time, attempt to realize a consequent standardization of product components in order to tap into the mass market early on. Local adaptations and changes regarding the language are therefore based on a standardized core product. This approach benefits from English as the lingua franca, which fosters fast expansion into global markets.

Through their strategy to reach economies of scale on a global level, the companies successfully set global "de facto" standards early on. This initiates a self-enforcing feedback loop from which the companies benefit (cf. "growth strategy"). The global standards are set in the US market and enforced in the international markets by the power of US-American companies' enormous financial resources (cf. "capital").

The analyzed Israeli and Scandinavian companies attempted to focus on the English-language version of their products early on. Their business models were targeted for international expansion right from the beginning. Entering the North American market is a central target, especially for Israeli companies. Part of the internationalization strategy was therefore to set up a local office in Silicon Valley, in order to identify new market trends, engage in networking, and hire qualified employees (cf. "ecosystem").

### *4.4.4  Recommendations*

The following recommendations can help companies to improve the scalability of their products and business models:

- Companies
  - Outsourcing and offshoring of non-core activities are conducive to focusing on a company's core business, preparing for internationalization, and increasing staff flexibility in times of growth or declining business. Specialization and internationalization often go hand in hand, especially in the software industry. Therefore, companies should consider outsourcing or offshoring non-essential business and service activities.
  - Cloud-based platform solutions play an increasingly crucial role both as scalable infrastructure and as the core of a company's product. Software-based companies, even to a larger extent than their customers, can benefit from using cloud platforms as a lever for increasing scalability. Especially in terms of offering their services more flexibly, companies should leverage the potential of cloud solutions in order to manage growth processes more easily and become increasingly location-independent.
  - Companies should consider two possible strategies to scale up their business: on the one hand, they should create a (distribution) platform (this applies especially to B2C companies). The advantage is that the companies themselves can act as a gatekeeper. On the other hand, companies can build upon a standard that has already been established (e.g. iOS respectively the App Store).

- Public Sector
  - Government funding of cloud solutions in the form of research and development funding should be evaluated, as they act as an incentive for companies to develop scalable solutions that can be easily internationalized.
  - Furthermore, in order to improve the international perspectives for German software-based companies, public sector support to establish European or global standards and interfaces would be advisable.
  - It is also recommended that Internet connectivity (both broadband and mobile access) is expanded and improved, in order to provide companies with the necessary infrastructure to scale up their product portfolio according to their needs.

**Best Practice: "The Art of Minimalism"**

Does this road lead to the future? Whoever decides to take the country road to get to this dynamic start-up in Kaiserslautern enjoys a picturesque countryside. The small road meanders past medieval castle ruins, through narrow valleys, and dense forests along the hills of Rhineland-Palatinate. After a thunderstorm, wafts of mist crawl up the hills. Can you really find high-tech in a place like this? If one continues just a few kilometers, you get pretty close to the spirit of Silicon Valley.

Several software start-ups have settled in the suburbs of Kaiserslautern and are expanding, some of them very successfully, to the global markets. A glass elevator takes you up to the second floor of the stylish entrepreneurship center where the smart company's headquarters are located. A dynamic atmosphere emanates from the employees. "Marketing is an essential part of our business strategy," explains the young company founder.

The man in his mid-thirties founded the company together with two fellow students at the end of the 1990s. The founders specialized early on in the development of mobile software solutions for companies. The company is active in several dozen countries. How did the company manage to expand internationally so quickly? **"We focus on our core business. We outsource everything that can be outsourced,"** the CEO tells us. The management cooperates with different partner companies in the global target markets, **giving them a share of product sales**, and thereby obtains valuable international contacts.

A **cloud-based platform solution greatly reduces the need for local sales and service staff**. Thanks to this strategy, the company can attain the **capability to scale up** quickly, which would be impossible to realize internally with staff-intensive service offerings. "I cannot imagine how other companies can aspire to grow quickly if they do not **concentrate on the core business**," diagnoses the founder.

To further drive the scalability and quick growth, the company, originally financed with a bank loan, gradually opened up to risk capital investors. **The additional capital can be invested in marketing and PR measures to make the company known internationally**. This is essential, especially for gaining new partners. Even more important, however, are the investors' international contacts in the local target markets. Through them, new partners and international clients can be attracted.

In addition, **specialists** who are in high demand worldwide **are attracted by being offered shares in the company**, thereby making specific know-how to the company available.

**Outsourcing wherever possible, scale up rapidly, and entering into partnerships**. The company from Kaiserslautern in the German countryside will most certainly expand into many other countries. After all, one travels more easily with less luggage and a local travel guide.

## 4.5 Growth Strategy

Along with the product strategy and scalability, the study also analyzed software-based companies' growth strategy. In this context, we also analyzed the impact of one specific institutional factor, the size of the domestic market, on a company's cross-country expansion.

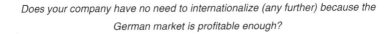

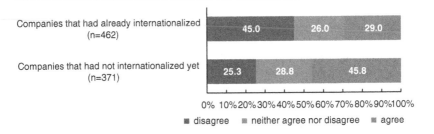

**Fig. 4.6** Descriptive statistics—perceived profitability of the domestic market

### 4.5.1 Quantitative Results

45.0 % of the surveyed companies that had already internationalized disagreed on the question of whether the German market is profitable enough[1] and they therefore do not need to internationalize any further; 29.0 % of them agreed, while 26.0 % of the companies neither agreed nor disagreed.

The findings are quite different for the companies that had not internationalized yet. 45.8 % of the surveyed companies in this group agreed that the German market is profitable enough and they therefore do not have to internationalize; this contrasts with only 25.3 % of the companies in this group that did not consider the market profitable enough and 28.8 % that neither agreed nor disagreed (cf. Fig. 4.6).

The regression analysis yields some interesting findings regarding the impact of the domestic market's perceived profitability: the analysis shows a significant impact on the degree of internationalization. Software-based companies that consider the German market attractive enough internationalize to a lesser extent—regardless of how successful the companies act objectively on the German market. The German market seems to be a *"comfort zone"* which the companies feel too comfortable to leave. The qualitative study could further substantiate and differentiate these results.

### 4.5.2 Qualitative Case Study Results—Germany

The German market turns out to be large enough for many companies in the software industry to do quite well. One can regard the German market as a kind of "comfort zone". Therefore, not all companies feel the need to internationalize quickly.

---

[1] In the quantitative survey, we focused specifically on the domestic market's perceived profitability, which our interviews revealed is closely related to the size of the domestic market.

> Germany is this interesting case, where [market size] is (...) in the middle. It's big enough (...) to be able to build a small medium sized technology company. But it's not as big as an American market place, or as China.
> (Venture capitalist, Silicon Valley)

However, due to the limited size of the German market, the majority of the analyzed companies cannot grow to become international leaders by only operating within the domestic market. German companies will face a fierce competition, especially in markets with strong network effects, once international competitors with a large number of customers enter the market.

The analyzed companies' international expansion was also characterized by the fact that the companies are relatively risk averse and consider the cost of internationalization in detail (cf. "capital"). The analyzed companies also tend to avoid the very competitive American market. The companies' growth is often opportunistic and lacks a long-term vision.

The companies often focus on business models that strongly depend on staff-intensive services, which impede both a fast expansion into international markets and scaling up the business quickly. In contrast, some of the analyzed companies deliberately made use of outsourcing solutions in order to keep the expenses for staff low and to be able to quickly expand internationally (cf. "scalability").

> Basically you have to decide earlier than you like. (...) You cannot do it half way. You either have to spend the effort upfront and say I want to be the best and then do it early or you are not going to do that at all. (...) We had to get there first so that someone could not just rip us off. The network effect in technology are always that the winner takes it all.
> (CTO and founder of an application software company, Silicon Valley)

The variety of cultures and different languages of the European market makes it difficult for companies to expand across borders into neighboring countries, regardless of efforts of regulatory and legal harmonization within Europe.

If on the other hand one considers companies that operate in niches in the B2B segment, one will find that due to the limited size of the respective segment, they are forced to internationalize early on. As the German market offers comparatively little lasting potential in specialized segments, these companies internationalize earlier. At the same time, competition in these highly specialized market segments is often less intense. This especially benefits companies financed by their founders, as they can follow a sustainable growth strategy more easily in specialized niches (cf. "capital").

> Laziness. I mean it's a lot more work (...) to understand markets outside Germany and especially to go international. And I think a lot of, lot of Germans just like (...) having a small, nice, comfortable life. (...) Founder attitudes have a huge impact here.
> (Professor, founder, and business angel, Stanford)

Considering the findings, the analysis shows that German software-based companies' international expansion, in contrast to the internationalization of companies from Silicon Valley, is mainly driven by their founders. Expansion thus strongly depends on the management's ambition and assertiveness.

### 4.5.3  Qualitative Case Study Results—International

In contrast to the German companies, the analyzed companies in Silicon Valley operate in a local market with stronger competition. This market offers higher opportunities and risks at the same time. The size of the American market is seen as a distinct competitive advantage. The large growth potential allows for fast national expansion of the companies that we studied. Therefore, once they internationalize their business, the companies have already reached a considerable size. As the US market presents such a large single market, companies face significantly less coordination costs than companies in Europe, where each market has to be targeted and entered separately.

US risk capital investors' expected rate of return demands for quick and offensive growth. It is important to note in this context that it is not only the founders of the analyzed companies in Silicon Valley that drive the growth strategy. Instead, many involved parties, which depend financially on the company's success, support the company's continuous expansion (cf. "cooperation partners" and "capital").

At the same time, the companies' growth strategy is characterized by the involvement of international experts, who are often recruited within the founders' social networks or from national and international competitors. These experts are usually responsible for the regional expansion abroad (cf. "cooperation partners").

It did not make a difference whether the companies decided to expand within the American market or internationally. The only factor that matters is fast expansion, which enables for the highest possible valuation, when selling the company after an average of 5 years (cf. "capital"). Considering the large amount of capital invested, crossing national borders is considered to be an obvious next step in the growth process.

Another positive driver of the growth strategy can be found in Silicon Valley's ambitious and success-oriented environment. This environment additionally encourages founders to further expand (cf. "ecosystem").

In contrast, the small domestic markets of Israel and Scandinavia forced the analyzed companies to expand internationally early on. As Israeli companies' options are limited in terms of entering neighboring markets, the case study partners often focus on global markets. As employees in Israeli and Scandinavian companies speak English very well and the companies are well connected with the USA, companies can enter the English-speaking market early on.

### 4.5.4  Recommendations

To optimize German software-based companies growth strategies, the following recommendation can be derived for companies and the public sector:

- Companies

    - Companies that target markets with strong network effects should consider early internationalization and, if necessary, consider the option to seek financing by risk capital investors.
    - When choosing risk capital investors, companies should not only focus on the investor's financial resources but also on its international business network.
    - If internationalization is a viable option for the company's business model, the founders should develop a focused international vision as a "think big strategy" right from the start and prepare the technical and managerial aspects of the internationalization early on.
    - If the team of founders does not want to consider external financing, it is often advisable to follow a niche strategy. In this approach, companies should attempt to penetrate a specific market segment as extensively as possible, in order to subsequently leverage this expertise and follow a cautious international expansion strategy.
    - The target market should be selected systematically and, possibly, by considering cultural proximity. In general, geographical proximity should not be a crucial factor.

- Public Sector

    - The public sector could tailor its procurement policies to small local IT companies, which would be beneficial for these companies by offering them a growth perspective early on.
    - We also recommend that the public is made aware of existing German software champions. Domestic software-based companies that have internationalized successfully could be praised and their achievements made public.

---

**Best Practice: "Piggyback from the Niche into the World"**

What does a true "hidden champion" look like in the software industry? The hidden champion does not only avoid the public spotlight, but actually hides, in the truest sense of the word. The headquarters of this hidden champion are similarly hidden in the outskirts of a small Munich community, surrounded by neat row houses with groomed yards, fields, and meadows. Unpretentious, but not without a well-justified self-confidence, the company sparkles within this suburban serenity. Having entered the spacious foyer, the visitor can see the glass-covered courtyard. The founder welcomes us, walking down a sweeping staircase. He founded the company together with two fellow students back in 1980, offering product-based consulting solutions in the B2B segment. The company now employs more than 4,000 employees and generates revenues of more than half a billion Euros—quite impressive compared to many other software-based companies in Germany. What is the secret of their success?

The sustainable growth model, based on a strategy of vertical depth, has proven successful in the last decades. This strategy is based on **the employees building up above-average expertise and penetrating a specific market segment as extensively as possible**. This strategy has two advantages: on the one side, through the **company's critical know-how**, products and complementary consulting solutions can be created that differ from the competition's offerings. On the other side, **international competition in small market segments is often less fierce**. Especially **companies that are financed by their founders** and follow a sustainable growth plan can expand steadily with this strategy. "Regarding our internationalization, we decided to only go abroad in the segments in which we were also successful in the domestic market," explains the founder. Once the company reaches a high degree of expert knowledge in one segment, **internationalization is planned in detail jointly with partner companies**.

Especially the company's **bigger clients, that already have internationalized, help in entering international markets**. By using their clients' physical and informational infrastructure, the management succeeds in internationalizing "piggyback" style. **Once a niche has been successfully entered in Germany or in an international market, the company subsequently attempts to enter further related segments**. This kind of persistent growth strategy has turned out to be very successful in the last decades.

The company will continue its **thoughtful expansion** and will further expand from the fields just outside Munich into large markets all over the world.

## 4.6 Cooperation Partners

Access to cooperation partners and international contacts is another factor that we investigated through quantitative analysis. In this context, we studied the relationship between management's lack of international contacts and the degree of internationalization of the surveyed companies.

### 4.6.1 Quantitative Results

For the empirical analysis of the impact of international business contacts, we split the surveyed companies into two groups. The first group had already entered international markets and generated some part of its revenue abroad. The second group was only operating in the domestic market. 44.0 % of the companies that had already internationalized did not agree that their management was missing the

*Is the management lacking international contacts for (further)
internationalization?*

Companies that had already internationalized (n=461): disagree 44.0, neither agree nor disagree 21.0, agree 34.9

Companies that had not internationalized yet (n=368): disagree 31.3, neither agree nor disagree 26.9, agree 41.8

0% 10% 20% 30% 40% 50% 60% 70% 80% 90% 100%

■ disagree        ■ neither agree nor disagree        ■ agree

**Fig. 4.7** Descriptive statistics—lack of management's international business contacts

necessary business contacts to further internationalize the company's business. In contrast, 34.9 % considered the missing international contacts an impediment to internationalization, while 21.0 % of the companies neither agreed nor disagreed.

In contrast, 41.8 % of the surveyed companies that had only been active in the German market indicated that their management was lacking international contacts in order to internationalize their business. 31.3 % of the companies disagreed, while 26.9 % neither agreed nor disagreed (cf. Fig. 4.7).

The regression analysis shows that lacking international contacts indeed have a significant negative impact on software-based companies' degree of internationalization. Thus, a lack of international contacts can be considered as one of the causes of a company's limited degree of internationalization. These results could be further substantiated and analyzed in more detail in the qualitative case studies.

## 4.6.2 Qualitative Case Study Results—Germany

The analysis of the German cases studies shows that the analyzed companies' step abroad was often taken when a favorable opportunity opened up. Business networks and local market knowledge are an essential part of the market entry strategies (cf. "scalability", "product strategy", and "growth strategy").

The founders purposefully draw on existing social ties to international business partners in order to enter target markets. The management's social environment can thus be regarded as the main source of international contacts. Decisions to internationalize are often taken in an opportunistic way and depend on contacts within the social and business network (cf. "growth strategy"). The analysis also shows, in this context, that small and medium-sized companies are often missing high-caliber international contacts, which makes internationalization more difficult. Based on our analysis, we could identify the following three strategies for entering international markets:

First, companies draw on contacts to local individuals and leverage these contacts in order to establish further contacts to business partners.

Second, companies follow their domestic clients that have already internationalized. The companies then work on projects in the clients' international offices and extend existing contracts.

> [We considered internationalizing] very quickly because of the demand of customers. It was nearly right away. Because it's an enterprise software. (…) When you deal with those large organizations you need to do that.
> (Founder of a company for talent management software, San Francisco)

Third, companies deliberately enter into cooperation with larger companies in order to use their international infrastructure. The partner companies provide the software-based companies with international contacts and a social infrastructure (cf. "scalability").

Especially the last two strategies turned out to be successful when entering new geographical markets. Metaphorically speaking, one could call this a "piggyback" strategy. Sitting on the shoulders of both clients and business partners, the companies cross borders and enter new markets.

Regarding B2C companies, it is mainly investors and local online communities that help to overcome the obstacles of internationalization. The companies benefit from their investors who provide the necessary market knowledge and access to customers. In some cases, this facilitates setting up sales offices in the respective markets (cf. "capital").

Formal networking events and informal contacts were important to all analyzed companies when hiring highly qualified employees, especially with a technology focus, and international sales staff. We would like to highlight that qualified staff was mainly recruited within social networks. Personal recommendations and references are another important factor (cf. "human factors").

### 4.6.3  Qualitative Case Study Results—International

Entering international markets via customers and partners is common practice for the analyzed software-based companies in Silicon Valley. By securing investment from external individuals, such as risk capital investors, serial entrepreneurs, lawyers, and consultants, the companies' management team can draw on a large group of experienced individuals, who in turn support the companies with their business networks. As they are invested in the companies, they have a direct incentive to share their experience and contacts with the companies (cf. "capital").

International market entry is only to a limited degree the result of a favorable opportunity, but rather the consequence of a planned growth process (cf. "growth strategy"). In addition, the companies benefit from the large number of big software companies, international founders, and employees that are located in this region. By tapping into the networks of individuals connected to the company, founders gain access to potential customers and employees with special skills.

The international origins of individuals from Silicon Valley allow the companies to start their internationalization process in their domestic market (cf. "ecosystem"). Contacts with employees from other geographical regions, clients, and business partners are usually established in Silicon Valley. Thus, entering international markets is only seen as an extension of local activities.

Foreign founders in Silicon Valley additionally use existing and trustworthy contacts with their home countries to drive their companies' internationalization into these countries. It is also common practice to establish flexible cooperation between start-ups and large international corporations. This opens up the opportunity, even for small companies, to be taken "piggyback" into global markets by targeting the large corporations' international clients.

Contacts with individuals that had emigrated were a crucial door opener for the analyzed Israeli and Scandinavian companies to enter international markets. Especially for Israeli companies, good connections with Silicon Valley and North America were helpful in entering the American market.

## 4.6.4 Recommendations

From the findings of our study, we can derive the following recommendations for companies and the public sector:

- Companies
  - Even when choosing the first members of the advisory board and risk capital investors, a major emphasis should be placed on their international experience and their business networks. Opportunities often arise from their networks whose value might not be seen in the company's earlier phases, but can turn out to be invaluable later on. The members of the advisory board and investors also need to be willing to take risks when the founders later require their approval for the internationalization of the company.
  - German software-based companies with an aggressive growth strategy should also consider opening an international office in Silicon Valley. In addition to obtaining international business contacts and getting closer to international customers, companies can acquire knowledge about current industry and market trends, as well as having the opportunity to hire exceptionally talented employees. If nothing else, this office might open up the opportunity to be financed by risk capital investors from Silicon Valley.
  - German software-based companies with a defensive strategy should consider a "piggyback" strategy for entering international markets. With this strategy, international markets can be entered either with German clients that have already internationalized or are currently expanding, or indirectly by using

their business network, e.g. working with a common sales service provider. This approach is especially helpful for B2B companies in order to gain references one country at a time with little risk.

- Public Sector
  - It is recommendable for the public sector to further support and extend existing incubators abroad (such as the "Silicon Valley Accelerator"). These institutions provide German software-based companies with a first contact point in the international markets and the option to use it as a starting point to extend their local networks. At the same time, incentives should be created for foreign incubators to increasingly open offices in Germany.
  - Furthermore, the exchange between young start-ups and large established international corporations should be fostered. Our analysis shows that easier access to customers that already have internationalized can facilitate the internationalization process considerably.
  - High staff turnover rates in a region influence the extension of social contacts and business networks as well as the flow of industry knowledge across company borders.

---

**Best Practice: "The Sound of Success"**

What does success sound like? Glorious, like Verdi's triumphal March? Or rather dramatic, like Wagner's Ride of the Valkyries? On the sixteenth floor of the sober office building in San Jose, the sound of success reminds one of a calm piano concert by Chopin—beautifully composed, yet calm without being obtrusive.

The company has rented some functional office space which it shares with another company. The welcoming assistant asks us to enter the founder's office. The India-born CEO with a level-headed charisma has already founded several companies in Silicon Valley.

The last company sold for a remarkable price of more than 1.5 billion dollars. Why did he found another company? "To leverage the skills that I have acquired to become even more successful." A typical serial entrepreneur whose motivation and drive to reach new heights is further encouraged by the ambitious social environment in San Francisco's Bay Area. In a rather humble tone and melodic Indian English accent, the shrewd founder talks about the concept for success of his latest masterpiece.

He discovered the security concerns of many customers in regard to cloud computing early on, and developed a technology that allows for intra-company encryption and storage of data. Since the company was founded in 2010, the company has won dozens of clients from many different countries. **A shortage of capital has never been a problem for the affluent founder,** as he had made enough money with his earlier companies. **Still, he decided early to get a venture capital fund invested in his company**.

Among the many possible investors, the founder picked one of the most renowned funds in Silicon Valley. His **objective** was not to raise additional capital, rather to **obtain valuable contacts**. Shortly thereafter, the company could successfully sign contracts with several international customers.

The company enters into **partnerships** with global consulting companies that sell the product to their customers. This way, new markets can be entered quickly and expenses for staff can be kept low. This strategy also does not require expensive offices abroad at the beginning. **Only once the company has won several clients in a country that require personal support, does the management consider opening a branch office**.

The management also focuses on extending the **business network**, in order to use the network for reaching out to new international clients and getting them interested about the product. In this context, an **efficient PR apparatus**, which increases consumer awareness of the company, is indispensable to sales success.

The founder's various **entrepreneurial experiences** make it possible to systematically develop the market. Likewise, it is easier for an experienced composer to write a new masterpiece. The audience is likely to continue to hear the melodic tune played by the company from San Jose in harmony with its experienced partners.

## 4.7 Ecosystem

While the previous sections have focused on the managerial options of the entrepreneur, this final section emphasizes the whole ecosystem. A software-based company's location is often considered essential to its success. In this context, the term "cluster" has increasingly received attention. Clusters have been defined as agglomerations of companies that operate in the same or related industries and are located in spatial proximity. From the study's results, we were able to shed light on the impact of cluster regions on software-based companies' degree of internationalization.

### 4.7.1 Quantitative Results

37.4 % of the surveyed companies indicated that their company is located in an "IT region". In contrast, 40.3 % of the companies are located in a geographical area that

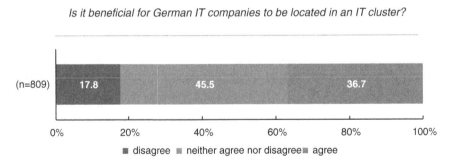

**Fig. 4.8** Descriptive statistics—perceived benefits of IT cluster regions

is not known as an IT region, and 22.3 % of the companies were undecided whether their company is located in an IT region.

36.7 % of the respondents agreed that having an IT company located in an IT cluster is beneficial for German companies. Only 17.8 % of the companies did not consider having a company located in a cluster region as an advantage. 45.5 % neither agreed nor disagreed on this question (cf. Fig. 4.8).

However, the regression analysis shows that a software-based company's location in a cluster region does not have a significant impact on a company's degree of internationalization. Thus, a company's location in a cluster cannot be seen as the cause of companies' international success. The results of the qualitative analysis shed important insight on the role of the ecosystem for software-based companies.

### 4.7.2 Qualitative Case Study Results—Germany

The results of the analysis show that companies are usually founded close to the founders' hometown. The crucial factor when selecting a company's location is based less on the opportunity to be based in a cluster, but primarily on the founders' personal and local networks, as well as their regional ties. This is also the reason why software-based companies' locations are spread out regionally.

Especially the German case study companies that are financed by their founders are to a lesser degree embedded in the regional ecosystem. This finding is true regardless of where the companies are located. The case studies therefore confirm that a company's location in a cluster does not automatically have a significant positive impact on software-based companies' internationalization.

However, the case studies provide some evidence that an entrepreneurial climate is gradually developing in some of the German software clusters, from which especially young companies in the B2C segment can benefit. By specifically approaching investors and potential employees in local "meeting spots", such as restaurants and cafés, the companies can successfully benefit from the social networks in a software cluster. Many interview partners named the city of Berlin as an example of an ecosystem in which this social exchange happens almost naturally.

The study's results underline primarily the importance of soft factors, such as an informational infrastructure and the possibility for personal networking. The companies can strategically use the many opportunities for networking in a cluster in order to establish cooperation with companies that have already internationalized (cf. "cooperation partners"). Particularly for B2C companies that want to grow quickly, being located in a metropolitan area allows for better access to a local pool of qualified and internationally minded employees (cf. "human resources").

Successful clusters seem to emerge primarily "bottom up" from a creative-entrepreneurial environment, such as in Berlin. The analysis of the case studies shows that "technocratic" clusters, that are in a sense imposed "top down" on a region, do not seem to be very promising. In contrast, successful, self-organizing clusters develop a dynamic of their own that cannot be easily anticipated. However, one has to keep in mind that many clusters in Germany are comparatively young and a creative environment might still be evolving.

### 4.7.3 Qualitative Case Study Results—International

In contrast to the German case study companies, the companies that we analyzed in Silicon Valley were specifically founded in this region. The decision for the best location was made rationally by the analyzed companies. The criteria essential to the companies' founders were the access to local networks, top talents, and investors.

A high density of relevant and mutually connected business partners (cf. "capital" and "growth strategy") allows for quick access to physical and immaterial resources and suitable employees (cf. "human resources"). As the actors involved are connected through a multifaceted risk capital system, a valuable ecosystem evolves that benefits everyone.

While the companies benefit from the available local resources, they also face an intense competition for employees, capital, and customers. However, the studied companies do not perceive this area of conflict as an inhibiting factor, rather as a positive incentive. At the same time, the companies depend on Silicon Valley as a location, in order to maintain and extend their business networks.

> There are really amazing engineers in UK and Germany, good technical universities. And a lot of them we've hired have moved here too. A lot of people love living in San Francisco. (...) We want the smartest people to be very happy.
> (CEO and founder of a company for security apps, San Francisco)

Because of the variety of international skilled staff, industry pundits, and international software-based companies, companies can plan for their internationalization while they are still in Silicon Valley. The cooperation between international employees and globally operating clients enables a "local internationalization".

The proximity to international research institutes and entrepreneurship centers was emphasized as another positive factor when considering where to locate a company, as it allows new innovations to be transferred into local companies' business models quickly (cf. "product strategy").

### 4.7.4 Recommendations

In order to create an ecosystem for the German software industry, we could develop the following recommendations for the public sector and software-based companies:

- Companies
  - To make better use and extend the potential of clusters, we recommend that companies actively use business networks, increase the reach of their networks, as well as share them with other members of the cluster. The companies should also increase their efforts to connect with potential cooperation partners in their cluster regions.

- Public Sector
  - It seems more promising to focus supporting measures on existing industry agglomerations than on developing technocratic clusters in regions that cannot develop and sustain a fertile ground for companies on their own.
  - Along similar lines, the public sector should increasingly support the self-organization of natural software clusters. Especially the support of an innovative environment in university settings offers the chance to create an ecosystem for the German software industry.
  - Similarly, trade and industry associations could develop and support industry-wide events for strengthening companies' business networks. Associations should not restrict themselves to bringing young companies together. Rather, it seems promising to promote a "healthy mix" of young and small with large and established companies. This heterogeneous mix can stimulate cooperation and trigger the development of products with a high international market potential.

**Best Practice: "California so Close"**

It is eight o'clock in the morning when the first surfers arrive at the beach. At the Rothschild Boulevard, some hundred meters from the beach promenade, the first cafés have started to open. Two Israeli soldiers are sitting on a bench on the boulevard's green median strip and are enjoying the early morning sun. It is only in the first hours of the day that Tel Aviv is so serene, before the booming town awakens.

A vibrant IT scene, with an international reputation, has evolved here in the city's southern center and the city's suburbs. With its bars, cafés, and art galleries, the atmosphere reminds one of Berlin's young Bohemian neighborhoods, combined with Silicon Valley's dynamic. In one of the city's suburbs, we are greeted by the young entrepreneur as warm-heartedly as an old friend returning after a long trip. The smart head of the company obviously enjoys his work. The casual sofas in the company's lounge, the young employees wearing colorful shorts, and the ease with which employees interact with each other all remind one of a Californian start-up.

Over a platform, the company offers web application for download to customers all over the world. The freemium strategy aims at quickly winning a large number of new international customers. Digital advertising, marketing, and PR with global media partners all play a crucial role in this setting.

Since their company's beginning in 2008, the founders were planning for their company's internationalization—also because the small Israeli market does not offer enough growth potential. The CEO talks enthusiastically about the possibilities that a platform-based business model offers to an Internet company. The digital distribution and payment channels, along with worldwide access make it possible to sell products far beyond the country's borders. Especially for Israeli companies, for which gaining access to business partners in Israel's neighboring countries is problematic, a purely Internet-based business model opens up completely new options.

Although the product can be sold directly and virtually via online channels, the choice of a location for the company is still important. Being located in Tel Aviv has contributed crucially to the company's international growth, as the city is a **gateway for doing business with Silicon Valley**. The founder has skillfully used this advantage to entice American risk capital investors about his company and for establishing contacts with advertising and marketing partners in Silicon Valley. Accordingly, the company opened an international branch office in California in order to monitor the latest trends and to better maintain the necessary business networks in the world's largest software market. Moreover, the company can tap into a large pool of highly qualified IT specialists. Hence, the company benefits from the two locations' available resources and from the opportunities that the Internet's direct sales channels offer.

In the age of digitalization, California is only a mouse click away from the beaches of Tel Aviv.

# Chapter 5
# Conclusion

The software industry is of huge importance for the competitiveness and efficiency of the German economy. Its economic leverage not only manifests as an important industry by itself but also through the interconnectedness with other industries. However, only a limited number of German companies have managed to grow globally and internationalize their business. In order to improve this situation, the two-and-a-half-year long research project "*German Software Champions—Status quo, success factors, perspectives*" identified successful strategies for software-based companies to leverage their potential for growth and internationalization and ascertained the supporting macro-economic conditions. The status quo, the study's methodology, and the main findings are summarized in this final report.

It was the objective of the first, primarily quantitative, phase to conduct a statistically sound analysis of the current status of the German software industry. It encompassed the analysis of the managerial drivers of growth and internationalization, as well as the macro-economic conditions within which software-based companies operate in Germany. Based on these findings, we could identify seven important topics for closer analysis in the second, qualitative, phase. These seven focal topics were analyzed in detail by conducting a comparative analysis of national and international regions.

By looking beyond the German horizon to Europe, North America, Israel, and Russia, we could compare the findings from these regions with the findings from Germany. Based on this comparison, we were not only able to identify basic patterns to explain the quantitative results, but also to identify success strategies for growth and internationalization, as well as derive recommendations for companies and the public sector.

Table 5.1 summarizes the most important recommendations in bullet-point form:

In light of the different conditions in each country, it does not seem promising to directly copy a strategy or an ecosystem such as Silicon Valley. Rather, a German approach should be tailored to the local setting. The hidden champions and their strategies that were outlined in this study showed that even German software-based companies can grow and internationalize successfully. We were able to derive both

© Springer International Publishing Switzerland 2015

A. Picot et al., *The Internationalization of German Software-based Companies*,
Progress in IS, DOI 10.1007/978-3-319-13548-9_5

**Table 5.1** The study's recommendations in summary

| Topic | Companies | Public sector |
|---|---|---|
| Capital | Be receptive to international risk capital investors: pay attention to the investors' business network | Support venture capital as an asset class and simplify investments from outside Germany |
| Human resources | Organizational culture and incentives: create flat hierarchies and offer stock options in order to hire and retain specialists | Additional interdisciplinary programs of study: offer the opportunity to gain qualifications, in addition to regular (under)graduate courses, in order to train experts with an entrepreneurial vision |
| Product strategy | Plan for internationalization right from the start: use english as the primary product language and provide for multi-language capability in the software | |
| Scalability | Implement platform strategies: develop own platforms or use international platforms for international distribution | Foster and accelerate standard setting: encourage European solutions that scale easily within the European Union |
| Growth strategy | ".com" domain first, physical location second: evaluate growth options in terms of early but inexpensive internationalization; set up an office in Silicon Valley later on, in order to gain access to US venture capital | Public sector as a customer: make procurement policies more flexible and tailor them to small and medium-sized software-based companies |
| Cooperation partners | B2B "piggyback" strategy: establish contacts with large companies early on and use their offices to expand internationally; B2C: focus and leverage business partners | Small and medium-sized companies meet multinational corporations: organize conferences and events to establish contact between German small and medium-sized software-based companies and multinational corporations |
| Ecosystem | Active use of the potential of cluster regions: expertise, international employees, and informal networks | Support of natural cluster regions: strengthen existing industry agglomerations and resource concentrations |

universal recommendations for software-based companies and specific recommendations in regards to the important differences between B2B and B2C business models.

The study's findings show that the German approach to internationalize in B2B markets is comparatively slower but can nevertheless be sustainable. Companies focus on specialized market segments and, over the years, accumulate special expertise. Stable processes in the domestic market can be subsequently leveraged for international market success, often "piggybacking" on a large domestic client. Successful German companies do not, by any means, lack innovativeness; their cautious growth, which is often internally financed, does not prevent them from internationalization. While this strategy can be considered successful from an individual entrepreneur's point of view, it is not sufficient from an economic perspective in helping Germany gain international leadership. German B2B companies with their focus on small niches capitalize sporadically on software characteristics, such as economies of scale and network effects. As long as this holds true, hardly any global market leaders, with the potential to act as a consolidator, will emerge within Germany.

There are companies in B2C markets which can be considered champions. Other companies in the end user segment could learn from them how to develop a successful growth and internationalization strategy. Surprisingly, it is in the segments of gaming and dating, both with a strong focus on marketing, that German companies have succeeded internationally. These companies have managed to build a platform in Germany and have scaled it up quickly. Many of these companies have efficiently centralized their management in their headquarters and reach an impressive size before any physical international offices become necessary. To follow such a strategy, many of the companies make use of freelancers, international service providers, and distribution platforms.

In this respect, this study's results are encouraging for software-based companies both in B2B and B2C markets. It is possible to reach international market leadership, and the public sector can be supportive in reaching this goal more easily. Regarding the financing of companies, simplifications and indirect support could increase investments in the German software industry. Another example is the area of education and advanced training, in which additional interdisciplinary programs of study could help in meeting the demand for software-savvy managers and software specialists with an entrepreneurial vision.

However, the companies themselves actually decide on their own fate. By opting for appropriate strategies, using opportunities to scale up their business, and integrating suitable cooperation partners at an early stage, companies can tap into the existing potential for growth and internationalization. In the end, the often quoted cultural difference between German and American companies does not represent an obstacle.

# Appendix

## Expert Interview Guideline, Phase I

I. Introduction:
1. Introductory round
2. Introduction of the project

- Joint project of IOM, WIM, and CDTM; sponsored by the Federal Ministry of Education and Research
- Phase 1: Germany as a location for software-based companies; identification of micro- and macro-economic conditions; analysis of the impediments to growth and internationalization for small and medium-sized companies
- Phase 2: International perspective and comparison of Germany with other countries
- Objective: develop political and managerial recommendations of how the German software industry can catch up with international leaders

3. Note: This interview is exploratory with open questions, in order to generate hypotheses which can then be tested in the further course of this study.

II. Interview:
1. Open part of the interview

- What are the reasons that German software-based companies show little growth compared to their international competitors?
- What are the reasons that German software-based companies show a low degree of internationalization compared to their international competitors?
- What makes the software industry unique compared to other industries?
- What is specific to the German software industry when compared internationally?
- What areas hold the biggest opportunities for the German software industry in the future?

© Springer International Publishing Switzerland 2015
A. Picot et al., *The Internationalization of German Software-based Companies*,
Progress in IS, DOI 10.1007/978-3-319-13548-9

2. Specific questions

- Which of the following factors do you consider crucial for growth and internationalization on each level?
- Where do German software-based companies have their strengths and weaknesses?

  (a) Company:
      Corporate strategy, corporate culture, characteristics of the entrepreneur(s)
  (b) Industry:
      Competitive forces and environment, market size and demand, related industries, supporting industries
  (c) Networks and contacts:
      Contacts, cooperation and associations
  (d) Physical location:
      Human resources, capital, infrastructure, quality of life
  (e) Society:
      Cultural factors, cultural impact on entrepreneurial factors
  (f) Government:
      Government regulations, bureaucracy, intellectual property, subsidies

3. Concluding Questions

- If you were a policy-maker, which decisions would you make and which factors do you consider especially critical?
- What is characteristic of Germany as a company location, compared to other countries? What are Germany's advantages and disadvantages?
- What was your motivation to start your own company?
- From your perspective, what is the biggest motivation for German software entrepreneurs to start a company?

## Case Study Interview Guideline, Phase II

| | |
|---|---|
| Interviewer: | |
| Department: | |
| Company: | |
| Interviewee: | |
| Position: | |
| With company since: | |
| Location: | |
| Date: | |
| Duration of Interview: | |

**Introduction**
**Permission for recording**

*Interview guideline part 1: questions for the management*

*1.1 Questions regarding the history of the company*

1.1.1 What is special about your company?

1.1.2 Please provide us with a short summary and the approximate dates from the company's history and the most important events regarding its internationalization

*1.2 Questions regarding the motivation and the decision process*

1.2.1 When was the first time you considered internationalizing your company's business?

1.2.2 Why did you internationalize?

☐ (Un-)attractiveness of the home market

☐ Attractiveness of the international market

☐ Competitive pressure

☐ Good opportunity

☐ Economic market leadership

☐ Intrinsic motivation (own aspirations)

☐ Cost savings

☐ Use economies of scale (recover investments)

1.2.3 Game 1:

Here you see 8 figures that represent 8 different groups of people. Please place the figures on the circles of the game board depending on the influence these groups had on your company's internationalization:

– The influence can be positive, encouraging internationalization, (green circles) or negative, discouraging internationalization (red circles)

– The closer to the middle the figures are placed, the stronger this group's influence. The further outside, the weaker the influence

– Please place the figures that represent groups with no influence outside the circles

☐ Equity investors

☐ Debt capital investors

☐ Employees

☐ (Co-)founders

☐ Managers

☐ Family and friends

☐ Customers

☐ Competitors

☐ Other (?)

(continued)

(continued)

| 1.2.4 What characteristics of these groups above supported the internationalization of your company? |
| --- |
| ☐ Knowledge of the international market |
| ☐ International/migration background |
| ☐ Experience abroad |
| ☐ Education/training |
| ☐ Specific abilities |
| 1.2.5 How/where did you get to know these persons? |
| 1.2.6 When is the right time for a company to internationalize? |
| ☐ Certain number of employees |
| ☐ Revenue threshold |
| ☐ Stable internal processes |
| ☐ Positive cash flow |
| 1.2.6 How did you handle the risk regarding your company's internationalization? |
| 1.2.7 How did you minimize the risk? |
| ☐ Risk minimization strategy |
| ☐ Internationalization without own international office/subsidiary |
| 1.2.8 How did your competitors' behavior give your company guidance regarding your choice of an internationalization strategy? |
| 1.2.9 Who were your entrepreneurial role models? |
| ☐ Company role models (who, why?) |
| ☐ Personal role models (personal environment, general) |
| *1.3 Questions on the first international market entrance* |
| 1.3.1 What were the selection criteria for your first international market? |
| ☐ Market size (revenue) |
| ☐ Competition |
| ☐ Customers |
| ☐ Technology |
| ☐ Supply of labor |
| ☐ Culture |
| ☐ Language |
| ☐ Contacts |
| 1.3.2 Through what channels did you enter the market? |
| ☐ Organic growth (own subsidiary/branch office) |
| ☐ Acquisition (purchase of a company) |
| ☐ Joint venture |
| ☐ Cooperation |
| ☐ Online/marketplaces |
| 1.3.3 How did you build up the required knowledge for the internationalization? |
| ☐ Education and training/university |
| ☐ Competitors |
| ☐ Own experience |
| ☐ New employees |
| ☐ Investors (business angels) with international experience |

(continued)

(continued)

*1.4 Company strategy*

1.4.1 What characteristics of your business model are especially suitable for internationalization?

1.4.2 What characteristics of your business model are poorly suited for internationalization?

1.4.3 To what extent have you standardized the process of internationalization?

1.4.4 What technologies do you use to minimize the cost of internationalization?

1.4.5 What parts of the value chain are performed abroad (production, sales)?

1.4.6 What parts of the value chain are you planning to digitalize/take online within the following 12 months?

1.4.7 At what speed are you pursuing your company's internationalization?

☐ Quickly/slowly (why quickly or slowly?)

☐ Country by country versus several countries at a time (why sequentially or concurrently?)

1.4.8 Is your strategy to outsource everything that does not belong to the core tasks/competencies of your company?

☐ Concentration on core tasks

☐ Has this approach been helpful for the company's internationalization?

*1.5 Company location*

1.5.1 How important is the location for your company, as a supplier of digital goods/technologies?

☐ Social environment (if yes: why?)

☐ Benefit from the location's reputation

☐ Infrastructure (if yes: why?)

☐ Capital (if yes: why?)

☐ Work force (if yes: why?)

Contacts/networks (if yes: why?)

*1.6 Summary*

1.6.1 All in all, what aspects worked especially well and what did not work at all when you think of your company's internationalization?

*Interview guideline part 2: questions for specific departments*

| | |
|---|---|
| Interviewer: | |
| Department: | |
| Interviewee: | |
| Position: | |
| With company since: | |
| Location: | |
| Date: | |
| Duration of interview: | |

**Introduction**
**Permission for recording**

*2.1 Manager human resources (HR)*

2.1.1 To what extent did you need to hire new employees for your company's internationalization?

☐ Within the home county

☐ Abroad

2.1.2 What role do international employees play for your company's internationalization?

☐ Contacts

2.1.3 How do you recruit new employees for your company's internationalization?

☐ Trade/employment fairs

☐ In cooperation with universities

☐ Other types of cooperation

☐ Chambers of foreign trade

☐ Personal contacts

☐ Online (job portals, linkedIn, etc.)

☐ Agencies ("E-lancing")

2.1.4 How important is the cost of labor for your company's decision to internationalize?

2.1.5 When you think of your international offices: do you employ people from these countries or from your home country?

2.1.6 How many job opening does your company currently have (worldwide)?

2.1.7 How many applications does your company receive for each open position? (worldwide)

2.1.8 Does your company suffer from a shortage of qualified employees? If yes: what kind and to what degree?

    ☐ For the following within home country/abroad:

        ○ Technical skills

        ○ Business and managerial skills

        ○ Experienced employees

2.1.9 If there is a shortage: what measures could you take to make your company more attractive as an employer?

☐ Financial incentives, non-financial incentives, job security

☐ Different location

☐ Corporate culture

☐ HsR marketing

2.1.10 If there is a shortage: does the shortage force your company to make increased use of outsourcing?

2.1.11 What internal trainings and courses does your company offer?

2.1.12 What is your company's rate of staff turnover?

2.1.13 What aspects should the government focus on to improve the long-term supply of qualified labor for your company?

*2.2 Manager Legal affairs*

2.2.1 To what degree did the countries' different laws have an impact on your company's selection of international markets?

☐ Protection of intellectual property

☐ Data privacy

☐ Labor laws

☐ Tax laws

2.2.2 How do you handle the different countries' local laws regarding your company's internationalization?

☐ Data privacy

2.2.3 To what extent do local data privacy laws in your home country make it necessary to take specific measures?

☐ Extent of legal advice

☐ Business model

2.2.4 Have you considered the legal aspects of cloud computing?

If yes: what aspects encourage and what aspects discourage your company to make use of cloud computing?

*2.3 Manager finance*

2.3.1 Did your company need additional funding (equity/venture capital or debt capital) for its internationalization?

2.3.2 Through which sources of capital does your company finance its internationalization?

☐ Equity (%)

○ Venture capital, business angels

○ Family, friends, founder(s)

☐ Debt capital (%)

○ Loans

☐ Profits and reserves

☐ Public sector aids/grants

2.3.3 How do you get in contact with investors?

☐ Friends/acquaintances

☐ Business partners

☐ Internet

☐ Networking events

2.3.4 To what extent do your companies' investors support your company beyond mere financing?

☐ Know-how

☐ Contacts abroad

☐ Pressure to internationalize

2.3.5 What aspects have been the most expensive regarding your company's internationalization?

(continued)

(continued)

☐ Marketing

☐ Product development

☐ Staff

☐ Cost for consulting/advice

☐ Infrastructure

*2.4 Product manager/manager product development*

2.4.1 Has your company developed its products with internationalization in mind?

2.4.2 What aspects of product design/development did you pay special attention to regarding international markets?

☐ Functionality

☐ Design

☐ Marketing

☐ Language

2.4.3 How did a new technology, infrastructure, or platform influence your company's internationalization strategy?

☐ Cloud computing/software as a service

☐ Social networks

☐ Smartphones

☐ Broadband

2.4.4 Did your company use technologies for its internationalization that had not been used by your competitors or had only been used by a few competitors?

2.4.5 Where does your company get the know-how for the development of its products?

☐ Cooperation with universities

☐ Cooperation with other companies

☐ Internal

☐ Recruitment of employees

2.4.6 To what extent are national and international customers involved in your company's product design?

*2.5 Manager sales*

2.5.1 What distribution channels do you use? In which countries and why?

☐ Publisher (only in case of gaming companies)

☐ Local distributors/sales partners

☐ E-commerce

☐ Within a corporate group

2.5.2 How do you select your sales partners?

2.5.3 How do you get in contact with these sales partners?

2.5.4 What role does the Internet play in your company as a sales channel?

☐ Share

☐ Growth

2.5.5 What products/services are difficult to sell over the Internet? Where does the Internet reach its limits?

2.5.6 How intensively do you use cloud services, cloud computing, or cloud storage for your company's internationalization?

*2.6 Manager marketing*

2.6.1 In general, how would you describe your international marketing strategy?

☐ Price

☐ Product

☐ Place(ment)

☐ Promotion

2.6.2 What does your international marketing strategy mainly focus on?

☐ Gain new customers

☐ Stable customer base

☐ Image

☐ Lock-in effects (description: making it difficult for customers to switch)

2.6.3 What advertising channels does your company use?

☐ Print

☐ Online

☐ TV

☐ Fairs and events

☐ Recommendations

2.6.4 Game 2:

How do you present your company within the home market?

How do your present your company abroad?

What role does the brand of the country or the region that your company is located in play for your company's products and corporate strategy?

(continued)

(continued)

| |
|---|
| •How do your company's customers perceive the regional origin of your company? |
| •How would you describe your company's regional branding? |
| •How important is the regional branding for your company? |
| •How do you rate the reputation/recognition of your country for the software industry? |
| 2.6.5 To what extent do you use different marketing strategies for different countries? |
| 2.6.6 Would a brand "Software Made in <*Home Country*>" be useful for your company? If yes: why? |

*Interview guideline part 3: general information on the company*

| | |
|---|---|
| Year of founding | |
| Total revenue | |
| Revenue abroad | |
| Profit | |
| Number of domestic employees | |
| Number of employees abroad | |

| Countries in which revenue is generated | Number of locations/subsidiaries |
|---|---|
| ... | ... |

# References

Amal, M., & Freitag Filho, A. R. (2010). Internationalization of small- and medium-sized enterprises: A multi case study. *European Business Review, 22*(6), 608–623.

Andersson, S., Gabrielsson, J., & Wictor, I. (2004). International activities in small firms: Examining factors influencing the internationalization and export growth of small firms. *Canadian Journal of Administrative Sciences, 21*(1), 22–34.

Autio, E., Sapienza, H. J., & Almeida, J. G. (2000). Effects of age at entry, knowledge intensity, and imitability on international growth. *Academy of Management Journal, 43*(5), 909–924.

Bell, J. (1995). The internationalization of small computer software firms: A further challenge to 'Stage' theories. *European Journal of Marketing, 29*(8), 60–75.

Bell, J., Crick, D., & Young, S. (2004). Small firm internationalization and business strategy: An exploratory study of 'Knowledge-intensive' and 'Traditional' manufacturing firms in the UK. *International Small Business Journal, 22*(1), 23–56.

Berg, B. L. (2006). *Qualitative research methods for the social sciences* (6th ed.). Boston, MA: Ally and Bacon.

Bertschek, I., Erdsiek, D., Köhler, C., Ohnemus, J., & Rammer, C. (2011). *Internationalisierung deutscher IKT-Unternehmen*. Mannheim, Deutschland: Zentrum für Europäische Wirtschafts--forschung. Retrieved May 19, 2014, from ftp://ftp.zew.de/pub/zew-docs/gutachten/IKT_Internationalisierung2011.pdf.

BITKOM. (2011). *Software-Exporte erreichen Rekordwert*. Retrieved May 19, 2014, from http://www.bitkom.org/files/documents/BITKOM_Presseinfo_Software_Export-Import_2010_-_16_03_2011.pdf.

BITKOM. (2013). *ITK-Marktzahlen*. Retrieved May 19, 2014, from http://www.bitkom.org/files/documents/BITKOM_ITK-Marktzahlen_Maerz_2013_Kurzfassung.pdf.

Broy, M., Jarke, M., Nagl, M., & Rombach, H. D. (2006). Manifest: Strategische Bedeutung des Software Engineering in Deutschland. *Informatik-Spektrum, 29*(3), 210–221.

Bryman, A. (2004). *Social research methods* (2nd ed.). Oxford, UK: Oxford University Press.

Buckley, P. J., & Casson, M. (1976). *The future of the multinational enterprise*. London, UK: Macmillan.

Buxmann, P., Diefenbach, H., & Hess, T. (2012). *The software industry: Economic principles, strategies, perspectives*. Berlin, Germany: Springer.

Calvet, A. L. (1981). A synthesis of foreign direct investment theories and theories of the multinational firm. *Journal of International Business Studies, 12*(1), 43–57.

Chetty, S. K., & Hamilton, R. T. (1993). Firm-level determinants of export performance: A meta-analysis. *International Marketing Review, 10*(3), 26–34.

Coviello, N., & Munro, H. (1997). Network relationships and the internationalization process of small software firms. *International Business Review, 6*(4), 361–386.

Cusumano, M. A. (2004). *The business of software: What every manager, programmer, and entrepreneur must know to thrive and survive in good times and bad*. New York, NY: Free Press.

© Springer International Publishing Switzerland 2015

A. Picot et al., *The Internationalization of German Software-based Companies*, Progress in IS, DOI 10.1007/978-3-319-13548-9

Dunning, J. H. (1973). The determinants of international production. *Oxford Economic Papers, 25* (3), 289–336.

Dunning, J. H. (1988). The eclectic paradigm of international production: A restatement and some possible extensions. *Journal of International Business Studies, 19*(1), 1–31.

Dunning, J. H. (2001). The eclectic (OLI) paradigm of international production: Past, present and future. *International Journal of the Economics of Business, 8*(2), 173–190.

Egeln, J., & Müller, B. (2012). *ZEW Gründungsreport Dezember 2012.* Mannheim, Germany: Zentrum für Europäische Wirtschaftsforschung GmbH (ZEW). Retrieved May 19, 2014, from http://ftp.zew.de/pub/zew-docs/grep/Grep0212.pdf.

European Commission. (2010). *Internationalisation of European SMEs.* Brussels, Belgium: Entrepreneurship unit, directorate-general for enterprise and industry, European commission. Retrieved May 19, 2014, from http://ec.europa.eu/enterprise/policies/sme/market-access/files/internationalisation_of_european_smes_final_en.pdf.

European Commission. (2014). *What is an SME?* Retrieved May 19, 2014, from http://ec.europa.eu/enterprise/policies/sme/facts-figures-analysis/sme-definition/.

Gabrielsson, M., & Kirpalani, V. H. M. (2004). Born globals: How to reach new business space rapidly. *International Business Review, 13*(5), 555–571.

Hellmann, T., & Puri, M. (2002). Venture capital and the professionalization of start-up firms: Empirical evidence. *The Journal of Finance, 57*(1), 169–197.

Holl, F.-L., Menzel, K., Morcinek, P., Mühlberg, J. T., Schäfer, I., & Schlüngel, H. (2006). Studie zum Innovationsverhalten deutscher Software-Entwicklungsunternehmen. In F.-L. Holl (Ed.), *Entwicklungen in den Informations- und Kommunikationstechnologien* (Vol. 2). Berlin, Germany: Eigenverlag. Retrieved May 19, 2014, from http://www.bmbf.de/pubRD/innovationsverhalten_sw-entwicklungsunternehmen.pdf.

Holtbrügge, D. (Ed.). (2003). *Management multinationaler Unternehmen: Festschrift zum 60. Geburtstag von Martin K. Welge.* Heidelberg, Germany: Physica-Verlag.

Iansiti, M., & Levien, R. (2004). *The keystone advantage: What the new dynamics of business ecosystems mean for strategy, innovation, and sustainability.* Boston, MA: Harvard Business Press.

Isenberg, D. J. (2008). The global entrepreneur. *Harvard Business Review, 86*(12), 107–111.

Jackson, D. J. (2012). *What is an Innovation Ecosystem?* Retrieved May 19, 2014, from http://www.urenio.org/wp-content/uploads/2011/05/What-is-an-Innovation-Ecosystem.pdf.

Johanson, J., & Mattsson, L.-G. (1988). Internationalization in industrial systems—a network approach. In N. Hood & J.-E. Vahlne (Eds.), *Strategies in global competition* (pp. 303–321). New York, NY: Croom Helm.

Johanson, J., & Vahlne, J.-E. (1977). The internationalization process of the firm—a model of knowledge development and increasing foreign market commitments. *Journal of International Business Studies, 8*(1), 23–32.

Johanson, J., & Vahlne, J.-E. (1990). The mechanism of internationalization. *International Marketing Review, 7*(4), 11–24.

Johnson, J. E. (2004). Factors influencing the early internationalization of high technology start-ups: US and UK Evidence. *Journal of International Entrepreneurship, 2*(1–2), 139–154.

Jones, M. V., Coviello, N., & Tang, Y. K. (2011). International entrepreneurship research (1989–2009): A domain ontology and thematic analysis. *Journal of Business Venturing, 26*(6), 632–659.

Jung, M., Unterberg, M., Heuer, K., & Bendig, M. (2009). *Neue Handlungspotenziale zur Erhöhung von Zahl und Qualität nachhaltiger Unternehmensgründungen in Deutschland.* Hamburg, Germany: EVERS & JUNG GmbH. Retrieved May 19, 2014, from http://www.existenzgruender.de/imperia/md/content/pdf/studien/abschlussbericht_handlspot.pdf.

Kappich, L. (1989). *Theorie der internationalen Unternehmungstätigkeit: Betrachtung der Grundformen des internationalen Engagements aus koordinationskostentheoretischer Perspektive.* Munich, Germany: VVF.

Knight, G., Servais, P., & Madsen, T. K. (2000). *The born global firm: Description and empirical investigation in Europe and the United States*. Chicago, IL: American Marketing Association.

Knight, G. A., & Cavusgil, S. T. (2004). Innovation, organizational capabilities, and the born-global firm. *Journal of International Business Studies., 35*(2), 124–141.

Kutschker, M., & Schmid, S. (2011). *Internationales management* (7th ed.). Munich, Germany: Oldenbourg.

Kurbel, K., & Nowakowski, K. (2012). *Globale Transformation der Software- und Services-Branche: Wo bleiben die deutschen Unternehmen?* Retrieved May 19, 2014, from http://www.econstor.eu/bitstream/10419/57872/1/715284444.pdf.

Lamnek, S. (2010). *Qualitative Sozialforschung* (5th ed.). Basel, Switzerland: Beltz.

Leimbach, T. (2010). *Software und IT-Dienstleistungen: Kernkompetenzen der Wissensgesellschaft Deutschland*. Karlsruhe, Germany: Fraunhofer ISI.

Lünendonk. (2011). *TOP 25 der Standard-Software-Unternehmen in Deutschland* 2010. Retrieved May 19, 2014, from http://luenendonk.de/wpcontent/uploads/2011/05/LUE_SSU_2011_f300511.pdf.

Loane, S., McNaughton, R. B., & Bell, J. (2004). The internationalization of internet-enabled entrepreneurial firms: Evidence from Europe and North America. *Canadian Journal of Administrative Science, 21*(1), 79–96.

McDougall, P. P. (1989). International versus domestic entrepreneurship: New venture strategic behavior and industry structure. *Journal of Business Venturing, 4*(6), 387–400.

Metzger, G., Heger, D., Höwer, D., & Licht, G. (2010). *High-Tech-Gründungen in Deutschland –Hemmnisse junger Unternehmen*. Mannheim, Germany: Zentrum für Europäische Wirtschaftsforschung (ZEW). Retrieved May 19, 2014, from http://ftp.zew.de/pub/zew-docs/gutachten/hightechgruendungen2_10.pdf.

Miesenbock, K. J. (1988). Small business and exporting: A literature review. *International Small Business Journal, 6*(2), 42–61.

Moen, Ø., Gavlen, M., & Endresen, I. (2004). Internationalization of small, computer software firms: Entry forms and market selection. *European Journal of Marketing, 38*(9–10), 1236–1251.

OECD. (2012). *OECD Internet Economy Outlook 2012*. OECD publishing. Retrieved May 19, 2014, from http://dx.doi.org/10.1787/9789264086463-en.

Ojala, A., & Tyrväinen, P. (2007). Business models and market entry mode choice of small software firms. *Journal of International Entrepreneurship, 4*(2–3), 69–81.

Onetti, A., Odorici, V., & Presutti, M. (2008). The Internationalization of Global Start-Ups: Understanding the Role of Serial Entrepreneurs. Retrieved May 19, 2014 from http://eco.uninsubria.it/dipeco/Quaderni/files/QF2008_2.pdf.

Oviatt, B. M., & McDougall, P. P. (1997). Challenges for internationalization process theory: The case of international new ventures. *Management International Review, 37*(2), 85–99.

Picot, A., & Neuburger, R. (2010). *Internet-Ökonomie—Baustein C 13* (2nd ed.). Oldenburg, Germany: Institut für ökonomische Bildung.

Porter, M. E. (1990). The competitive advantage of nations. *Harvard Business Review, 68*(2), 73–91.

Ruokonen, M., Nummela, N., Puumalainen, K., & Saarenketo, S. (2008). Market orientation and internationalisation in small software firms. *European Journal of Marketing, 42*(11–12), 1294–1315.

Scheer, A.-W. (2001). *Start-ups are Easy, But....* Berlin, Germany: Springer.

Sharma, D. D., & Blomstermo, A. (2003). The internationalization process of born globals: A network view. *International Business Review, 12*(6), 739–753.

Söndermann, M. (2010). Monitoring zu ausgewählten wirtschaftlichen Eckdaten der Kultur- und Kreativwirtschaft 2010. Berlin, Germany: Bundesministerium für Wirtschaft und Technologie (BMWi). Retrieved May 19, 2014, from http://langzeitarchivierung.bibbvb.de/wayback/20121126113405/ http://www.miz.org/artikel/monitoring-wirtschaftliche-eckdaten.pdf.

Statistisches Bundesamt. (2008). *Gliederung der Klassifikation der Wirtschaftszweige.* Retrieved May 19, 2014, from https://www.destatis.de/DE/Methoden/Klassifikationen/GueterWirtschaftk lassifikationen/klassifikationenwz2008.pdf?__blob=publicationFile.

Stiehler, A., Böhmann, T., & Weber, M. (2009). *IT Services Made in Germany—Stärken, Erfolgsbeispiele und Strategien deutscher IT-Dienstleister im internationalen Wettbewerb.* Berlin, Germany: BITKOM. Retrieved May 19, 2014, from http://www.bitkom.org/files/documents/IT_Services_Made_in_Germany.pdf.

Strauss, A., & Corbin, J. (1990). *Basics of qualitative research: Grounded theory procedures and techniques* (2nd ed.). Thousand Oaks, CA: Sage.

Top 100 Research Foundation. (2011). *Global Software Top 100.* Retrieved May 19, 2014, from http://www.softwaretop100.org/global-software-top-100-edition-2011.

TruffleCapital. (2012). *Ranking of the Top 100 European Software Vendors.* Retrieved May 19, 2014, from http://www.truffle100.com/downloads/2012/TruffleEurope-2012-v9.pdf.

Vernon, R. (1966). International investment and international trade in the product cycle. *The Quarterly Journal of Economics, 80*(2), 190–207.

Zahra, S. A., Ireland, R. D., & Hitt, M. A. (2000). International expansion by new venture firms: International diversity, mode of market entry, technological learning, and performance. *Academy of Management Journal, 43*(5), 925–950.

Zerdick, A., Picot, A., Schrape, K., Artopé, A., Goldhammer, K., Lange, U. T., et al. (2000). *E-Conomics—strategies for the digital marketplace.* Berlin, Germany: Springer.

Printed in the United States
By Bookmasters